# Introduction to Algebra

## with Business Technology Emphasis

A problem solving approach to introductory algebra.

A competency-based course in workbook
form that includes problems from:

Accounting
Business
Computer Science
Corrections
Culinary Arts
EcoTourism
Hotel Restaurant Management
Management
Management Information Science
Management Information Systems
Office Management
Retail Management
Police Science
& Travel and Tourism

# R.R. Barrows & J.B. Hart

# Introduction to Algebra
## with Business Technology Emphasis

*Second Edition*

**R.R. Barrows**

**J.B. Hart**
*Hocking College*

**KENDALL/HUNT PUBLISHING COMPANY**
4050 Westmark Drive    Dubuque, Iowa 52002

Book Team

President and Chief Executive Officer    Mark C. Falb
Vice President, Director of National Book Program    Alfred C. Grisanti
Editorial Development Supervisor    Georgia Botsford
Developmental Editor    Tina Bower
Vice President, Production Editorial    Ruth A. Burlage
Production Manager    Jo Wiegand
Production Editor    Lynne Burlage
Design Manager    Jodi Splinter
Cover Designer    Suzanne Millius
Senior Vice President, College Division    Thomas W. Gantz
Vice President and National Field Manager    Brian Johnson
Senior Managing Editor, College Field    Paul Carty
Associate Managing Editor, College Field    Chad Chandlee
Associate Editor, College Field    Jason Ward

# Dedication

To my best friend, Stephanie.

Also, to my husband, Steven, my daughter, Bailey Jo, and my parents,
Edward and Sheryl Threm.

R.R. Barrows

This work is dedicated to R.J. Hart and L.H. Hart...both of whom have
contributed immensely to the quality of my life.

I would like to thank all of my colleagues and students who
were in anyway involved in this project.

J.B. Hart

# Table of Contents

# Preface

This competency based, problem solving approach to introductory algebra was developed for use in both the two year college setting and the high school technology preparation (tech-prep) setting. It was designed to engage and involve the student through the use of large numbers of technology relevant example problems and its workbook format. With important concepts and topics revisited and reinforced throughout, it should promote understanding and retention. Containing many times the number of exercise problems included in most introductory algebra texts, it provides the student with ample opportunity for practice and it virtually eliminates the need for instructor generated supplements. Since it includes some step by step solutions within the complete answer key, the text should facilitate out of class work for the student. Taken together, the features of this text and its workbook format should help to ensure that the student is demonstrating competency... particularly if the instructor insists that the student complete at least the odd numbered exercises in the workbook before taking a test.

To facilitate our colleagues whose busy schedules allow little time for course development, we offer multiple version chapter tests, final exams, and answer keys. Please inquire at 740-753-3591, extension 2267, or write to the address given below. Please feel free to send any criticisms or suggestions for improvement to:

R.R. Barrows & J.B. Hart
School of Arts and Sciences
Hocking College
3301 Hocking Parkway
Nelsonville, Ohio 45764-9704

# CHAPTER 1
## Measurement Systems and Unit Analysis

Introduction:

In previous courses we were introduced to the two primary systems of measurement: the metric or international system (SI) and the customary or United States system (US). In those courses we learned what is meant by each of the terms denominate number, unit, prefix, and auxiliary unit. We also learned that a conversion is a unit change between the two systems and that a reduction is a unit change within one of the systems. We were told that the unit change process, be it conversion or reduction, is called unit analysis. We learned how to do unit analysis.

In this chapter we will again study unit analysis. Our work begins with a review of the two primary measurement systems. Our second task will be to work refresher exercises in ordinary unit analysis. After that we will define and learn to work with what are called derived units. Our final area of study and one of the major goals of this chapter is to combine unit analysis with certain business concepts.

Objectives:

Upon completion of this chapter you will be able to:
1. Explain what is meant by each of the terms: denominate number, unit, base unit, derived unit, auxiliary unit, conversion, and reduction.
2. Compare and contrast the customary system of measurement with the metric system of measurement.
3. Change any unit of measurement to another unit (like quantity).
4. Use the unit change processes to solve relevant problems from business, computer science, corrections, culinary arts, hotel/restaurant management, travel and tourism, office management, and police science as well as everyday life.

## 1.1 Review of Measurement Systems

A measurement is essentially a comparison. Some attribute of an object is compared to an established standard. For example, when we say that the weight of a package is 4 pounds we mean that we have compared the package's weight to a standard known as the pound and that we have found the package to be 4 times as heavy as the standard.

Numbers (such as 4 pounds) that include both a numerical value and a unit of measurement are called *denominate numbers*. We encounter denominate numbers frequently in our daily lives. Four ounces of ground beef, one gallon of milk, three quarts of paint, 30 square yards of carpet, and 125 miles are each familiar examples. Denominate numbers play a role in business, computer science, corrections, culinary arts, hotel/restaurant management, travel and tourism, office management, and public safety services. These numbers help us to organize our work and to communicate our results. When we write 10 grams, our readers know that we mean 10 grams and not 10 pounds, 10 quarts, or 10 seconds.

Units most likely originated as an attempt to standardize trade quantities or distances. The establishment of units such as the pound and the gallon provided uniform standards for the sale of commodities like meat and milk, while the establishment of units such as the foot and the mile provided length standards for construction and travel.

A *system of units* is a set of specified units of length, mass, temperature, and time from which most all other units are derived. Throughout history there have been thousands of different systems used, but only two major systems have survived and are in use today...*the metric or international system* (SI) and the *customary* or *United States system* (US). For most quantities these systems use different units. For example, the base unit of length in the metric system is the meter whereas the base unit of length in the customary system is the foot. One exception is time... both systems use the second as the base time unit.

The metric system was developed in the last quarter of the eighteenth century. Originally its developers intended that the base units be defined in terms of actual physical quantities. The base unit of length, the meter, was to be 1/10,000,000 of the distance from the North Pole to the equator. The base unit of mass, the gram, was to be the mass of the water contained in a cube that measures one centimeter (1 centimeter = 1/100 of a meter) on all sides. However, because of very small errors made in surveying the distance from the North Pole to the equator, the values of the meter and gram do not agree with their original definitions. Now the meter is defined by an atomic standard... the meter is the length of 1,650,763.73 wavelengths of the red-orange light emitted from an atom of Krypton-86 under certain specified conditions. Also, due to the small size of the gram, it is now the kilogram (1 kilogram = 1,000 grams) that serves as the mass standard.

The customary system is much older than the metric system and is based on less precise standards than is the metric system. Once widely used by most English-speaking nations, the customary system has been abandoned by almost all countries except the United States. Ironically, the metric system was legally adopted by the United States Congress in 1866, but its use has never been made compulsory due to reluctance on the part of some US residents. This reluctance to switch to the metric system will be discussed later in this chapter section.

Each of these systems begins with *base quantities* and their corresponding *base units*. The base quantities common to both systems are length, mass, temperature, and time. These base quantities and their units are shown in Table 1.1.

## Table 1.1  Base Quantities and Units

| Quantity | Units | |
| --- | --- | --- |
| | Metric | Customary |
| 1. Length | meter (m) | foot (ft) |
| 2. Mass | kilogram (kg) | pound (lb) |
| 3. Temperature | degree Celsius ($1^o$ C) | degree Fahrenheit ($1^o$ F) |
| 4. Time | second (sec) | second (sec) |

Throughout the years many auxiliary units were added to the customary system. The foot family of units includes the inch, the yard, and the mile. The pound family of units includes the ounce and the ton. These auxiliary units were created to eliminate the forced use of fractions or large numbers. Having these auxiliary units allows us to write 0.75 foot as 9 inches, 528,000 feet as 100 miles, 0.25 pound as 4 ounces, and 820,000 pounds as 410 tons.

The metric system was designed differently. Rather than create auxiliary units the SI system uses prefixes that represent powers of ten. Table 1.2 contains a partial list of the metric prefixes, their values, and their symbols.

Like the auxiliary units of the customary system, the metric prefixes were created so that we may avoid the forced use of fractions or large numbers. For example, the prefix milli- (one thousandth) allows an engineer to specify a bolt diameter as 12 millimeters rather than 0.012 meters.

It is not unusual to see metric prefixes used with customary units. Many employment advertisements contain salary ranges in kilodollars such as 25 k to 32 k. A 10 megaton nuclear bomb is one with the destructive power of 10,000,000 tons of TNT. A 7 megabuck lottery jackpot is one with a $ 7,000,000 prize. While some mathematicians disagree with the practice of mixing metric prefixes with customary units, the general public does not seem to find these mixed system terms objectionable. In this text we shall allow the use of mixed system terms.

Yet another use for metric prefixes is the use of the terms kilobyte and megabyte by computer scientists. A byte is 8 bits or about the amount of storage space required to store a letter of the alphabet or a numerical digit.

## Table 1.2  Some Metric Prefixes , Their Values, and Their Symbols

### FRACTIONS

| PREFIX | VALUE | POWER OF 10 | SYMBOL |
|--------|-------|-------------|--------|
| deci-  | 1/10  | $10^{-1}$   | d      |
| centi- | 1/100 | $10^{-2}$   | c      |
| milli- | 1/1,000 | $10^{-3}$ | m      |
| micro- | 1/1,000,000 | $10^{-6}$ | μ   |

### MULTIPLES

| PREFIX | VALUE | POWER OF 10 | SYMBOL |
|--------|-------|-------------|--------|
| deka-  | 10    | $10^1$      | da     |
| hecto- | 100   | $10^2$      | h      |
| kilo-  | 1,000 | $10^3$      | k      |
| mega-  | 1,000,000 | $10^6$  | M      |
| giga-  | 1,000,000,000 | $10^9$ | G   |

Mentioned earlier in this chapter section was the reluctance of some segments of the US population to switch to the metric system. Many mathematicians feel that this is the result of a lack of everyday familiarity with metric quantities. Most US residents have a mental picture of the foot, the gallon, and the pound... Imagine a foot long hot dog, a gallon of milk, or a pound of hamburger... the picture is there. Let's learn some physical models so that we may develop the same familiarity for metric units. The meter is about the distance from the left shoulder of an adult male to the tip of his outstretched right hand or a little longer than one yard. The centimeter (1 centimeter = 1/100 meter) is about the width of the fingernail on an adult index finger. The millimeter (1 millimeter = 1/1,000 meter) is about the diameter of the wire from which paper clips are made. The kilogram is slightly more than 2.2 pounds... so a 1 kg bag of pretzels would contain a little more than 2.2 lb of pretzels. A liter is about 6% larger than a quart... so a 2-liter bottle of cola contains a little more than 2 quarts.

# Exercises  1.1

Write your answers in the spaces provided. If the item requires calculation then you are to show all the work in the space provided. The answers may be found in the rear of the book.

1.  What is a denominate number? Give five examples of denominate numbers each with a different unit.

4

2.  Discuss the likely origin of units. Why was the establishment of units so important for business and industry?

3.  Explain why the customary system includes auxiliary units. Your explanation must include examples which support your answer.

4.  Explain how denominate numbers help people who work in these fields communicate with each other: a) business, b) computer science, c) corrections d) culinary arts, e) police science, f) office management, g) hotel/restaurant management.

5.  Explain what is meant by this statement: The essence of measurement is comparison.

6.  Which system of measurement is older ... the customary system or metric system?

7.  Using the physical models discussed in the last paragraph of this chapter section, estimate each of the following quantities.

    a)  The thickness of a dime.

    b)  The diameter of a dime.

    c)  The diameter of a quarter.

8. Using the physical models discussed in the last paragraph of this chapter section, tell which quantity is greater.

   a) Six feet or 2 meters?

   b) Eight pounds or 4 kilograms?

   c) Three centimeters or 1 inch?

   d) Fifty pounds or 20 kilograms?

   e) Thirty-two fluid ounces or 1 liter?

   f) Thirty-two ounces or 1 kilogram?


In exercises 9 to 36 you are to use the values found in Table 1.2 to tell how many of each quantity there is.
**Example**: The liter is a unit of volume in the SI system. What fraction of a liter is 10 mL ?
Answer: The prefix milli means 1/1000 so we have 10( 1/1000 )L = (10/1000)L = (1/100) L = 0.01 L.
   Please keep in mind that you are NOT required to memorize units other than the four base units. You will not have to tell what a watt or a joule is! However, you must be able to work with them in the prefix sense.

9. How many dollars are in 25 kilodollars?


10. The watt (W) is a SI unit of power. For example, most microwave ovens use between 750 watts and 2,000 watts of power. How many watts in 25 kW? Is this a large wattage compared to a microwave oven?


11. The liter (L) is a unit of volume in the SI system. How many liters are in 250 kL?


12. What fraction of a liter (L) is 100 mL?

13. The ampere (A) is the base unit of electric current in both systems. An electric range/oven draws as much as 30 to 40 amperes depending on how many burners are in use and the temperature at which the oven is set. What fraction of an amp is 200 mA?

14. The joule (J) is a unit of energy in the metric system. For example, cardiac defibrillators used by physicians and paramedics to shock a patient's heart into a normal rhythm discharge about 200 to 400 joules of energy. How many joules are contained in 450 kJ?

15. The gram (g) is a unit of mass in the international system. How many grams are contained in 1.2 kg?

16. How many liters (L) are contained in 3.5 ML?

17. The calorie (Cal) is an energy unit in the metric system. It is also related to the dietary calorie. How many calories are contained in 750 kCal?

18. What fraction of a meter (m) is 800 mm?

19. What multiple of a gram (g) is 0.24 kg?

20. What fraction of a liter (L) is 370 mL?

21. What multiple of a meter (m) is 250 cm?

22. What fraction of a liter (L) is 550 mL?

23. What fraction of a second (sec) is 300 msec?

24. What fraction of a second is 250,000 μsec (microseconds)?

25. How many tons are contained in 25 kton?

26. How many tons are contained in 50 Mton?

27. How many meters (m) are contained in 1.7 Mm?

28. What multiple of a meter (m) is 2.5 hm?

29. The candela (cd) is the base unit of light intensity in the metric system. How many candela are contained in 1.2 Mcd?

30. How many grams (g) are contained in 3,500 mg?

31. What fraction of a meter (m) is 0.001 dam?

32. How many watts (W) are contained in 0.5 MW?

33. How many grams (g) are contained in 0.5 Mg?

34. How many meters are contained in 9.5 Mm?


35. How many meters are contained in 2.3 Mm?


36. What multiple of a meter is 3.5 dam (dekameter)?

## 1.2 Refresher Unit Analysis Exercises

In the preceding section we learned that a denominate number is a number that answers two questions: "How many?" and "of what? ". We also learned that there are two major systems of units... the customary or US system with its auxiliary units and the metric or SI system with its prefixes. Sometimes when we are working with denominate numbers it is necessary to change from one unit to another unit. *A change from one system to the other is called a conversion. A change within a system is called a reduction.* Collectively, conversion and reduction are called *unit analysis.* To convert or reduce a denominate number we simply multiply by 1.

### Example 1
Reduce 13 feet to inches.

We know that 12 inches = 1 foot. If we divide both sides of this equation by 1 foot we obtain:

$$\frac{12 \text{ in}}{1 \text{ ft}} = \frac{1 \text{ ft}}{1 \text{ ft}} = 1$$

Now multiplying 13 feet by 1 in the form of 12 in/1 ft yields:

$$13 \text{ ft} \bullet 12 \text{ in/ft} = 156 \text{ in}$$

The two ft units cancel because 1 ft/1 ft = 1. After this cancellation all that remains is:

$$13 \bullet 12 \text{ in} = 156 \text{ in}$$

### Example 2
A certain air carrier allows a passenger to carry on a maximum of 10 kilograms of luggage. Convert 10 kilograms to pounds.

Consulting Table 1.3 we find that 2.205 lb = 1 kg. Dividing both sides of the equation by 1 kg we obtain:

$$\frac{2.205 \text{ lb}}{1 \text{ kg}} = \frac{1 \text{ kg}}{1 \text{ kg}} = 1$$

Multiplying:

$$10 \text{ kg} \bullet \frac{2.205 \text{ lb}}{1 \text{ kg}} = 22.05 \text{ lb}$$

Note: It is not necessary to divide both sides of the defining equation (2.205 lb = 1 kg in this case). Once we realize that we are taking advantage of cancellation we may substitute directly into the multiplication, letting cancellation be our guide.

# Table 1.3   Reduction Factors and Conversion Factors

## Length

| Metric Reduction Factors | Customary Reduction Factors | Conversion Factors |
|---|---|---|
| 1 Mm = 1,000,000 m | 1 ft = 12 in | 1 in = 2.54 cm  (exact) |
| 1 km = 1,000 m | 1 yd = 3 ft | 1 ft = 30.48 cm |
| 1 hm = 100 m | 1 mi = 1,760 yd | 1 m = 39.37 in |
| 1 dam = 10 m | 1 mi = 5,280 ft | 1 m = 3.281 ft |
| 1 m = 10 dm | | 1 mi = 1.609 km |
| 1 m = 100 cm | | |
| 1 m = 1,000 mm | | |
| 1 m = 1,000,000 $\mu$m | | |
| 1 dm = 10 cm | | |
| 1 cm = 10 mm | | |
| 1 mm = 1,000 $\mu$m | | |

## Mass/ Weight

| Metric Reduction Factors | Customary Reduction Factors | Conversion Factors |
|---|---|---|
| 1 kg = 1,000 g | 1 oz = 8 dram | 1 kg = 2.205 lb |
| 1 hg = 100 g | 1 lb = 16 oz | 1 lb = 453.6 g |
| 1 dag = 10 g | 1 ton = 2,000 lb | 1 t = 1.103 ton |
| 1 g = 10 dg | | |
| 1 g = 100 cg | | |
| 1 g = 1,000 mg | | |
| 1 g = 1,000,000 $\mu$g | | |
| 1 t (metric ton) = 1,000 kg | | |

## Time

| | | |
|---|---|---|
| 1 min = 60 sec | 1 hr = 60 min | 1 hr = 3,600 sec |
| 1 day = 24 hr | 1 week = 7 day | 1 yr = 365 day |
| 1 century = 100 yr | 1 millennium = 1,000 yr | |

## Area

| Metric Reduction Factors | Customary Reduction Factors | Conversion Factors |
|---|---|---|
| 1 m$^2$ = 1,000,000 mm$^2$ | 1 ft$^2$ = 144 in$^2$ | 1 in$^2$ = 6.45 cm$^2$ |
| 1 m$^2$ = 10,000 cm$^2$ | 1 yd$^2$ = 9 ft$^2$ | 1 m$^2$ = 10.76 ft$^2$ |
| 1 cm$^2$ = 100 mm$^2$ | 1 mi$^2$ = 27,878,400 ft$^2$ | 1 acre = 4,047 m$^2$ |
| | 1 acre = 43,560 ft$^2$ | |

| Metric Reduction Factors | 640 acre = 1 mi$^2$ | |
|---|---|---|
| 1 are = 100 m$^2$ | | Conversion Factors |
| 1 hectare = 100 are | | 1 hectare = 2.471 acre |

11

### Volume

| Metric Reduction Factors | Customary Reduction Factors | Conversion Factors |
|---|---|---|
| $1\ m^3 = 1{,}000{,}000{,}000\ mm^3$ | $1\ ft^3 = 1{,}728\ in^3$ | $1\ in^3 = 16.39\ cm^3$ |
| $1\ m^3 = 1{,}000{,}000\ cm^3$ (cc) | $1\ yd^3 = 27\ ft^3$ | $1\ m^3 = 35.31\ ft^3$ |

### Capacity ( special volume units without the cubic unit )

| Metric Reduction Factors | Customary Reduction Factors | Conversion Factors |
|---|---|---|
| 1 mL = 1 cc | 1Tbsp = 3 tsp | 1 fl oz = 29.57 mL (cc) |
| 1 L = 1,000 mL = 1,000 cc | 1 cup = 16 Tbsp | 1 L = 1.057 qt |
| 1 L = 100 cL | 1 pt = 2 cup | 1 gal = 3.785 L |
| 1 L = 10 dL | 1 qt = 2 pt | |
| 1 cL = 10 mL | 1 gal = 4 qt | |
| 1 dL = 10 cL | 1 barrel = 31.5 gal | |
| | 1 qt = 32 fl oz | |
| | 1 bushel = 9.309 gal | |

### Volume/Capacity

| Metric Reduction Factors | Customary Reduction Factors | |
|---|---|---|
| $1\ m^3 = 1{,}000\ L = 1\ kL$ | 1 board foot =144 $in^3$ | 1 gal = 231 $in^3$ |
| $1\ cm^3 = 1\ cc = 1\ mL$ | $1\ ft^3 = 7.481$ gal | 1 cord = 128 $ft^3$ |
| | 1 bushel = 1.244 $ft^3$ | |

## Example 3

Reduce 72 inches to feet.

Using 1 ft = 12 in from Table 1.3:

$$72\ in \bullet \frac{1\ ft}{12\ in} = 6\ ft$$

## Example 4

Reduce 75 gallons to quarts.

Using 1 gallon = 4 quarts from Table 1.3:

$$75\ gal \bullet \frac{4\ qt}{1\ gal} = 300\ qt$$

## Example 5

Convert 125 centimeters to inches.

Using 2.54 centimeters = 1 inch from Table 1.3:

$$125 \text{ cm} \bullet \frac{1 \text{ in}}{2.54 \text{ cm}} \approx 49.2 \text{ in}$$

Sometimes we may wish to use what are called *compound denominate numbers.* Seven feet 3 inches, 6 pounds 4 ounces, and 12 meters 5 centimeters are all examples of compound denominate numbers.

## Example 6

Reduce 31.25 feet to a compound denominate number using feet and inches.

Since the number is already expressed in feet we need only to reduce the fractional portion from feet to inches.

$$0.25 \text{ ft} \bullet \frac{12 \text{ in}}{1 \text{ ft}} = 3 \text{ in}$$

Thus we have 31 ft 3 in.

## Example 7

A police officer measured a set of skidmarks at an accident site and found them to be 34.5 yards long. Reduce 34.50 yards to a compound denominate number using feet and inches.

Using 3 feet = 1 yard from Table 1.3:

$$34.5 \text{ yd} \bullet \frac{3 \text{ ft}}{1 \text{ yd}} = 103.5 \text{ ft}$$

$$\text{Then } 0.50 \text{ ft} \bullet \frac{12 \text{ in}}{1 \text{ ft}} = 6 \text{ in}$$

Thus we have 103 ft 6 in.

## Example 8

Reduce 30.55 meters to a compound denominate number using meters and centimeters.

$$0.55 \text{ m} \bullet \frac{100 \text{ cm}}{1 \text{ m}} = 55 \text{ cm}$$

So we have 30 m 55 cm.

## Example 9

A technician measured the volume of a solution and found it to be 1,700 milliliters. Convert this volume to quarts.

When we view Table 1.3 we are unable to find an equation relating milliliters and quarts. In situations such as this we are forced to find an intermediate unit to act as a stepping stone. Searching Table 1.3 we find the equations: 1 L = 1,000 mL and 1 L = 1.057 qt. With these two equations we can perform our unit change in two steps.

Step 1: Reduce milliliters to liters.
Using 1,000 mL = 1 L from Table 1.3:

$$1,700 \text{ mL} \bullet \frac{1 \text{ L}}{1000 \text{ mL}} = 1.7 \text{ L}$$

Step 2: Convert liters to quarts.
Using 1.057 qt = 1 L:

$$1.7 \text{ L} \bullet \frac{1.057 \text{ qt}}{1 \text{ L}} \approx 1.8 \text{ qt}$$

## Example 10

Mt. Clarista, a popular tourist attraction, is 7,230 ft tall. Convert 7,230 feet to kilometers.

Step 1: Reduce feet to miles.
Using 5,280 ft = 1 mi:

$$7,230 \text{ ft} \bullet \frac{1 \text{ mi}}{5,280 \text{ ft}} \approx 1.37 \text{ mi}$$

Step 2: Convert miles to kilometers.
Using 1.609 km = 1 mi:

$$1.37 \text{ mi} \bullet \frac{1.609 \text{ km}}{1 \text{ mi}} \approx 2.20 \text{ km}$$

Note: It is not necessary to perform these calculations in discrete steps. We may treat multi-step problems as chain multiplications.

$$7,230 \text{ ft} \bullet \frac{1 \text{ mi}}{5,280 \text{ ft}} \bullet \frac{1.609 \text{ km}}{1 \text{ mi}} \approx 2.20 \text{ km}$$

## Example 11

A recipe for soup calls for 1 gallon 3 quarts of chicken broth. Convert 1 gallon, 3 quarts to liters.

Step 1: Reduce gallons and quarts to quarts.

$$1 \text{ gal} \cdot \frac{4 \text{ qt}}{1 \text{ gal}} + 3 \text{ qt} = 4 \text{ qt} + 3 \text{ qt} = 7 \text{ qt}$$

Step 2:  Convert quarts to liters.

$$7 \text{ qt} \cdot \frac{1 \text{ L}}{1.057 \text{ qt}} \approx 6.6 \text{ L}$$

Unit analysis can be used to do currency exchange calculations as is shown in the next two examples.

**Example 12**
Change $150.00 in U.S. dollars to British pounds if today's exchange rate is $1.55 = 1£.

$$\$150.00 \cdot \frac{1£}{\$1.55} \approx 96.8 \text{ £}$$

**Example 13**
A travel agent wishes to exchange $1,200 U.S. dollars for urkels. If today's exchange rate is 2.4 urkels = $1.00, how many urkels should the travel agent recieve?

$$\$1200.00 \cdot \frac{2.4 \text{ urkels}}{\$1.00} = 2,880 \text{ urkels}$$

# Exercises  1.2

Show all work in the spaces provided. The answers may be found in the rear of the book. Round answers to hundredths if rounding is needed.

1.    Reduce 28 ft (feet) to in (inches).


2.    Reduce 70 oz (ounces) to lb (pounds).


3.    Convert 30 mi (miles) to km (kilometers).


4.    Reduce 775 mm (millimeters) to cm (centimeters).

5.    Convert 500 lb (pounds) to kg (kilograms).

6.    Convert 40 qt (quarts) to L (liters).

7.    Reduce 30,500 cm (centimeters) to m (meters).

8.    Write 17.125 lb (pounds) as a compound denominate number using lb (pounds) and oz (ounces).

9.    Express 10,750 ft (feet) as a compound denominate number using mi (miles) and ft.

10.   Write 712 m (meters) 46 cm (centimeters) as a decimal denominate number with meters as its only unit.

11.   Express 105 cm (centimeters) 2 mm (millimeters) as a decimal denominate number using cm as its only unit.

12.   Reduce 64 oz (ounces) to pounds.

13. Reduce 2.4 dm (decimeters) to cm (centimeters).

14. Reduce 1.2 mm (millimeters) to μm (micrometers).

15. Convert 10 tons to t (metric tons).

16. Reduce 25 hm (hectometers) to dam (dekameters).

17. Convert 600 km (kilometers) to miles.

18. Reduce 300 μg (micrograms) to mg (milligrams).

19. Convert 0.80 lb (pounds) to g (grams).

20. Reduce 4 Tbsp (tablespoons) to tsp (teaspoons).

21.  Reduce 2.5 gal (gallons) to pt (pints).

22.  Reduce 40 gal (gallons) to bushels.

23.  Reduce 80 gal (gallons) to barrels.

24.  Reduce 5.2 barrels to gal (gallons).

25.  Reduce 0.25 L (liters) to mL (milliliters).

26.  Reduce 5 mL (milliliters) to cc (cubic centimeters).

27.  Reduce 0.45 L (liters) to cc (cubic centimeters).

28.  Reduce 3.25 hr (hours) to sec (seconds).

29.  Reduce 720 hr (hours) to min (minutes).

30.  Reduce 220 min (minutes) to hr (hours).

31.  Reduce 0.25 day to sec (seconds).

32.  Reduce 1.75 days to min (minutes).

33.  Convert 1,240 mm (millimeters) to in (inches).

34.  Convert 4.7 in (inches) to mm (millimeters).

35.  Convert 80.4 gal (gallons) to L (liters).

36.  Reduce 8,000 oz (ounces) to tons.

37.  Convert 72 in (inches) to m (meters) .

38.  Convert 1,800 mL (milliliters) to gal (gallons).

39.  Reduce 2,350 yd (yards) to mi (miles).

40.  Convert 2,900 yd (yards) to km (kilometers).

41.  Convert 12 tons to kg (kilograms).

42.  Convert 250 mL (milliliters) to fl oz (fluid ounces).

43.  Reduce 1,000 kton (kilotons) to Mton (megatons).

44.  Reduce 44 yd (yards) to in ( inches).

45. Convert 40 oz (ounces) to kg (kilograms).

46. Convert 3.5 kip (kilopounds) to kg (kilograms).

47. Reduce 5,000 kg (kilograms) to Mg (megagrams).

48. Convert 700 mi (miles) to kilometers.

49. Write 20.25 lb as a compound denominate number using lb and oz.

50. Express 20,270 feet as a compound denominate number using mile and feet.

51. Write 630 m 25 cm as a decimal denominate number with meters as its only unit.

52. The cargo on a certain semi-truck has a mass of 17,379 kg (kilograms). Convert this mass to tons.

53. A chef needs 4 pounds of beef for a recipe. How many kilograms does she need?

54. A dishwasher detergent is sold by a European company in 200 L (liter) drums. Convert this volume to gallons (gal).

55. A certain project on which your employees are working requires 7 mm (millimeter) diameter bolts. Your parts department carries only bolts with diameters measured in inches (in). Convert 7 mm to inches.

56. An oil truck tank has a volume of 3.6 kL (kiloliters). Convert this volume to gallons (gal).

57. An airline limits your total luggage to 40 kilograms. How many pounds is this?

58. The legal cargo weight for a semi-trailer in your state is 40,000 lb. How many a) tons is this? b) kilograms is this?

59. Your agency rents cars. A customer drove a rental car 425 miles. How many kilometers did she drive?

60. An airline passenger received 1,200 frequent flyer miles. How many kilometers is this?

61. If the currency exchange rate today is 2.5 Deutsch Marks = $1.00 U.S. Change 1,700 DM to U.S. dollars.

62. Change 7,500 £ (British Pound) to U.S. dollars if today's exchange rate is 1£ = $1.55.

63. A standard height for desks and computer work tables is 28 inches. How many centimeters is this?

64. A police officer measured a set of skidmarks at an auto accident and found them to be 78 ft, 3 inches long. The insurance company report asks for the length in meters. How many meters is 78 ft 3 inches?

## 1.3 Derived Units

A *derived unit* is any unit created using multiplication, division, or multiplication and division of two or more base units. Our study of derived units begins with area, proceeds to volume, and concludes with fractional units such as pressures, densities, speeds, and other rates.

## Area

Consider the square shown in Figure 1. It measures 3 ft by 3 ft or 1 yd by 1 yd. If we calculate its area or coverage using both of the units we find that:

3 ft x 3 ft = 1 yd x 1 yd or  9 ft² = 1 yd²

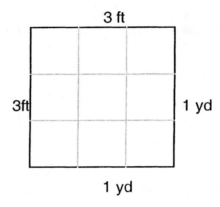

Figure 1

## Example 1
Reduce 640 square yards to square feet.

Using 9 ft² = 1 yd² from above:

$$640 \text{ yd}^2 \bullet \frac{9 \text{ ft}^2}{1 \text{ yd}^2} = 5{,}760 \text{ ft}^2$$

## Example 2
Convert 7,250 square feet to square meters.

We need to make an area equation that relates square feet to square meters. Using 3.281 ft = 1 m from Table 1.3, we have:

$$(1 \text{ m})^2 = (3.281 \text{ ft})^2 = 10.76 \text{ ft}^2$$

Then:

$$7{,}250 \text{ ft}^2 \bullet \frac{1 \text{ m}^2}{10.76 \text{ ft}^2} \approx 674 \text{ m}^2 \quad \text{(rounded to ones)}$$

## Example 3

A hotel parking lot measures 85 yards by 320 yards. Find its area in acres. The area of a rectangle is given by A = l • w, where l = length and w = width.

$$A = 320 \text{ yd} \bullet 85 \text{ yd} = 27,200 \text{ yd}^2$$

Since Table 1.3 gives 43,560 ft2 = 1 acre, we must reduce 27,200 yd2 to ft2

$$27,200 \text{ yd}^2 \bullet \frac{9 \text{ ft}^2}{1 \text{ yd}^2} = 244,800 \text{ ft}^2 \text{ (using the full calculator panel )}$$

Reducing to acres:

$$244,800 \text{ ft}^2 \bullet \frac{1 \text{ acre}}{43,560 \text{ ft}^2} = 5.619834711 \text{ acres which rounds}$$

to 5.6 acres if we round to the tenths place.

Note:   We could have reduced yards to feet before calculating the area.

## Volume

Consider the cube shown in Figure 2. It measures 3 ft by 3 ft by 3 ft or 1 yd by 1 yd by 1 yd. If we calculate the volume or space occupied using both of the units we obtain 3 ft x 3 ft x3 ft = 27 ft3, or 1 yd x 1 yd x 1 yd = 1 yd3.

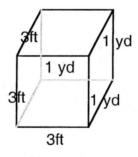

Figure 2

So it takes 27 cubic feet to fill up 1 cubic yard.  Other volume equations may be obtained by cubing both sides of the defining length equation.

## Example 4

Reduce 720 cubic feet to cubic yards.

$$720 \text{ ft}^3 \bullet \frac{1 \text{ yd}^3}{27 \text{ ft}^3} \approx 27 \text{ yd}^3 \text{ (rounded to the ones place)}$$

## Example 5
Reduce 3,850 cubic inches to cubic feet.
Since 1 ft = 12 in, we have:

$$(1 \text{ ft})^3 = (12 \text{ in})^3 = 1,728 \text{ in}^3$$

Then:

$$3,850 \text{ in}^3 \cdot \frac{1 \text{ ft}^3}{1,728 \text{ in}^3} \approx 2.23 \text{ ft}^3 \ (\text{ to hundredths })$$

## Example 6
A storefront in a strip mall is rectangular in shape and its interior measures 28.0 feet by 60.0 feet. Find the volume of the store if it has ceilings that are 10.0 feet high.

The volume of a rectangular prism (box) is given by V= l • w • h where l = length, w = width, and h = height.

So V = 60.0 ft • 28.0 ft • 10.0 ft = 16,800 ft³

Many special volume units were established so that we could refer to volumes without using cubic units. These include the barrel, the gallon, the fluid ounce, the liter, and the milliliter. See Table 1.3 for these special volumes and how they relate to cubic units.

## Example 7
Reduce 16.3 cubic feet to gallons.

Using 7.481 gal = 1 ft³ from Table 1.3:

$$16.3 \text{ ft}^3 \cdot \frac{7.481 \text{ gal}}{1 \text{ ft}^3} \approx 122 \text{ gal}$$

## Example 8
Convert 1.2 gallons to cubic centimeters.

Step 1: Reduce gallons to quarts.
Step 2: Convert quarts to liters.
Step 3: Reduce liters to milliliters. (1 mL = 1 cc)

$$1.2 \text{ gal} \cdot \frac{4 \text{ qt}}{1 \text{ gal}} \cdot \frac{1 \text{ L}}{1.057 \text{ qt}} \cdot \frac{1000 \text{ mL}}{1 \text{ L}} \approx 4,541 \text{ cc}$$

## Example 9
A shipping crate measures 60.0 in by 45.0 in by 30.0 in. Find its volume in cubic feet.

$$V = l \bullet w \bullet h = 60.0 \text{ in} \times 45.0 \text{ in} \times 30.0 \text{ in} = 81,000 \text{ in}^3$$

Then:

$$81,000 \text{ in}^3 \bullet \frac{1 \text{ ft}^3}{1,728 \text{ in}^3} = 46.875 \text{ ft}^3 \approx 47 \text{ ft}^3$$

## Fractional Units

Most derived units will be fractional units. These include rates of change, densities, pressures, and many others. Fractional units are used very heavily in business, corrections, culinary arts, police science, hotel/restaurant management and office management. These include copies per minute, parts per minute, dollars per hour, output per shift, words per minute, cents per mile, crimes per year, inmates per corrections officer, dollars per inmate, inmates per cell, inmates per square foot, rooms per unit, ounces per portion, and calories per ounce. Other than having the extra denominator unit or units to change, the reduction and conversion processes are the same. As usual, we shall let cancellation be our guide.

## Example 10
Change $9.00 per hour to dollars per work week.

$$\frac{\$9.00}{\text{hr}} \bullet \frac{40 \text{ hr}}{\text{week}} = \frac{\$360.00}{\text{week}}$$

## Example 11
Change 1.2 auto thefts per day to auto thefts per year.

$$\frac{1.2 \text{ auto thefts}}{\text{day}} \bullet \frac{365 \text{ day}}{\text{year}} = \frac{438 \text{ auto thefts}}{\text{year}}$$

## Example 12
Change $4.00 per pound to cents per ounce.

$$\frac{\$4.00}{\text{lb}} \bullet \frac{100¢}{\$1.00} \bullet \frac{1 \text{ lb}}{16 \text{ oz}} = \frac{25¢}{\text{oz}}$$

## Example 13

A hotel contains 80 rooms which rent for $57.00 per day per room. If the hotel was fully occupied for an eight day convention, find the amount of revenue for those eight days. Note: $57.00 per day per room gives us a complex fraction... that is, a fraction within a fraction.

$$80 \text{ rooms} \bullet \frac{\dfrac{\$57.00}{\text{day}}}{\text{room}} \bullet 8 \text{ days} = \$36,480$$

## Example 14

On a qualifying exam a state trooper is asked to change 60 miles per hour to feet per second.

Step 1:  Reduce miles to feet.
Step 2:  Reduce hours to seconds.

$$60 \ \frac{\text{mi}}{\text{hr}} \bullet \frac{5,280 \text{ ft}}{1 \text{ mi}} \bullet \frac{1 \text{ hr}}{60 \text{ min}} \bullet \frac{1 \text{ min}}{60 \text{ sec}} = 88 \ \frac{\text{ft}}{\text{sec}}$$

Right now you may be asking this question: I am not a police science student, so why do I have to do problems like Example 14? The answer is very simple. If you can do problems like Example 14 you can do similar rate problems from your technology. Examples 15 through 20 demonstrate such rate problems.

## Example 15

A production line at a manufacturing plant has two presses that produce 250 parts per hour and 200 parts per hour respectively. How long will it take both presses working together to produce 9,000 parts?

The two presses working together have a combined production rate of 450 parts per hour. Then changing 9,000 parts to hours:

$$9,000 \text{ parts} \bullet \frac{1 \text{ hr}}{450 \text{ parts}} = 20 \text{ hours}$$

## Example 16

A new laser printer has a print rate of 6 pages per minute. How long will it take the printer to print a 224 page document?

$$224 \text{ pages} \bullet \frac{1 \text{ minute}}{6 \text{ pages}} = 37.\overline{3} \text{ minutes}$$

## Example 17

A restaurant uses a high volume city waterline that delivers 20.5 gpm (20.5 gallons per minute). Convert the flow rate 20.5 gpm to liters per hour.

Step 1: Convert gallons to liters.
Step 2: Reduce minutes to hours.

$$20.5 \frac{gal}{min} \cdot \frac{4\ qt}{1\ gal} \cdot \frac{1\ L}{1.057\ qt} \cdot \frac{60\ min}{1\ hr} \approx 4,654.7 \frac{L}{hr}$$

## Example 18

How long would it take three 150.0 gpm filter pumps to filter a hotel swimming pool that is estimated to hold 33,500 cubic feet of water?

Step 1: Reduce cubic feet to gallons.
Step 2: Change gallons to minutes.

$$33,500\ ft^3 \cdot \frac{7.481\ gal}{1\ ft^3} \cdot \frac{1\ min}{450.0\ gal} \approx 557\ min$$

We use 450.0 gpm since three 150.0 gpm pumps working together are equivalent to one 450.0 gpm pump. Since 557 min is a rather large number of minutes, we reduce to hours:

$$557\ min \cdot \frac{1\ hr}{60\ min} \approx 9.28\ hr$$

## Example 19

Fans and blowers are rated by the volume of air they move over a specified time period. For example, a 1,200 cfm fan moves 1,200 cubic feet of air per minute. If the state environmental health agency requires that the air in a restaurant be changed every 10 minutes, find the minimum cfm rating for a fan in a restaurant that measures 120.0 feet by 40.0 feet by 12.0 feet.

$$V = l \cdot w \cdot h = lwh = 120.0\ ft \cdot 40.0\ ft \cdot 12.0\ ft = 57,600\ ft^3$$

$$\frac{57,600\ ft^3}{10\ min} = 5,760 \frac{ft^3}{min} = 5,760\ cfm$$

So we need at least a 5,760 cfm fan.

# Exercises 1.3

Show all work in the spaces provided. The answers may be found in the rear of the book. Round answers to hundredths if rounding is needed.

1.  Reduce 11.5 ft$^2$ to in$^2$.

2.  Reduce 175 cm$^2$ to mm$^2$.

3.  Convert 300.0 cm$^2$ to in$^2$.

4.  Convert 105 m$^2$ to ft$^2$.

5.  Reduce 255 yd$^2$ to ft$^2$.

6.  A proposed lot for a restaurant is rectangular and measures 200 ft by 750 ft. How many acres does it cover?

7.  The outer walls of a hotel measure 720.0 ft by 10.0 ft (disregarding windows). Given that one gallon of the specified paint covers 350 ft$^2$, how many gallons of paint are required to give the hotel two coats of paint?

8.  A restaurant has three dining rooms that measure 36 ft by 48 ft, 16 ft by 32 ft, and 12 ft by 24 ft. If the carpet the manager selected costs $11.25 per yd$^2$ including padding and installation, find the cost of carpeting those three rooms.

9.  A restaurant kitchen floor is to be covered with linoleum. If the kitchen measures 36 ft by 24 ft how many square yards of linoleum are needed?

10. An office that measures 16 ft by 20 ft is to be carpeted on the floor and four feet up the walls. How many square yards of carpet are needed?

11. A hotel manager has three housekeeping employees who are going to seal the hotel parking lot with asphalt topping. If one five-gallon bucket of the topping covers 50 yd$^2$ and the rectangular parking lot measures 200.0 ft by 600.0 ft, how many buckets of topping should the manager order?

12. A rectangular lot that was purchased to build an office complex has an area of 7.8 acres. If its width is 450 ft, what is its length?

13. Reduce 7,000 in$^3$ to ft$^3$.

14. Reduce 0.38 L to cm$^3$ (cc).

15. Reduce 2,750 cm$^3$ to L.

16. General Motors' best selling V8 engine to police departments is the 350 cubic inch engine. Convert 350 in$^3$ to liters.

17. Reduce 1,000 gallons to ft$^3$.

18. Convert 475 m$^3$ to ft$^3$.

19. Convert 12.5 ft$^3$ to m$^3$.

20. Convert 2.2 fl oz to cm$^3$ (cc).

21.   Convert 750 mL to fl oz.

22.   A tank shaped like a rectangular prism measures 10.0 ft by 4.0 ft by 3.0 ft. Find the tank's volume in gallons.

23.   A certain four cylinder engine has a displacement of 2.2 liters. Convert 2.2 liters to $in^3$.

24.   A flat bottom swimming pool at a hotel measures 20.0 m by 10.0 m by 1.5 m. Find the pool's volume in a) $m^3$, b) kL, and c) gal.

25.   A filtering system on a swimming pool has a 80.0 gpm (80.0 gallons per minute) filter pump. If the pool holds 1,200 $ft^3$ of water. How long will it take the pump to circulate all the water through the filter?

26.   Your company just bought a lot for a restaurant and it has an obsolete 20,000 gallon underground storage tank. If the tank is to be filled with sand and the sand and gravel company's dump trucks can haul 336 $ft^3$ of dry sand each, how many truckloads of sand are needed?

27.   An air exchange system in a hotel conference room has a 2,500 cfm ($ft^3$/min) blower that moves the air. If the room measures 48.0 ft by 24.0 ft by 12.0 ft, how long will it take to exchange all the air in the room?

28.   Convert 60 mi/hr to km/hr.

29.   Convert 45 ft/sec to km/sec.

30.   Convert 75 m/sec to ft/min.

31. The standard issue 9 mm caliber handgun issued by many police departments has a bullet muzzle velocity of approximately 1,400 ft/sec. Reduce this speed to mi/hr.

32. In the United States we measure automobile fuel efficiency in miles per gallon of gasoline consumed. Convert 27 mi/gal to km/L.

33. Which car is more fuel efficient...car A at 30 mi/gal or car B at 13.5 km/L? By how many km/L?

34. Standard tire pressure is 32 pounds per square inch. Convert the pressure 32 lb/in$^2$ to kg/m$^2$.

35. Convert the pressure 50.0 kg/m$^2$ to lb/in$^2$.

36. An ink jet printer has a print rate of 4 pages per minute. How long will it take the printer to print a 122 page document?

37. A certain major metropolitan area reported a 1997 burglary rate of 1,337. Change 1,337 burglaries per year to burglaries per week

38. If the crime statistics for a certain city show 12 assualts per week, find the number of assualts per year.

39. A salesperson for your company drove 770 kilometers one week. If she is to be reimbursed 30¢ per mile for her travel, how much should she be paid?

40. If a 2 ounce portion of pudding contains 540 calories, how many calories are contained in 3.5 ounces of the pudding?

41. If a 3 ounce portion of sherbert has 690 calories, how many calories are contained in three pounds of the sherbert?

42. Two photocopiers are being used to reproduce 10-page flyers. If one prints 8 pages per minute and the other prints 12 pages per minute, how long would it take to print 350 of the flyers?

43. In a certain city the police department reports that the average full-time patrol officer handles 2.2 domestic violence calls per week. If the department has 32 full-time patrol officers, how many domestic violence calls are handled per year?

44. A medical secretary and her new assistant can process 10 insurance claims and 6 insurance claims respectively when working together. How many hours will it take them to process 200 claims?

45. An office manager for a group of physicians sees that new patient records folders are being created at the average rate of 30 per week. If she allows 1/4 inch per folder and each shelf unit has five-32 inch shelves, how many new shelf units will she need to accomodate records for the next 24 months?

46. A production line at a manufacturing plant has three molding machines that produce 90 pieces per hour, 120 pieces per hour, and 200 pieces per hour respectively. How long will it take all three presses working together to produce 8,200 pieces?

47. A fax machine has a receive and print rate of 8 pages per minute. How long will it take to receive and print a 21 page document?

48. A hotel contains 110 rooms which rent for $120.00 per day per room. If the average occupancy rate for the hotel is 70%, find the average monthly revenue for the hotel.

49. If the federal courts mandate that each inmate have 60 square feet of cell area and the state department of corrections houses 30 inmates in 72 feet by 24 feet dormitories, is the state meeting the federal mandate for area?

50. If the state estimates that it costs $31,050.00 per inmate per year, find the cost of housing 1,750 inmates for three months.

51. The state requires that there be one corrections officer per 24 inmates per shift. How many corrections officers are needed to staff a facility that has 1,750 inmates? (Remember there are three shifts per day and seven days per week).

52. The National Institute for Justice has issued standards for handcuffs. A regulation pair of handcuffs cannot weigh more than 425 g or exceed 240 mm in length. The handcuffs must also be able to withstand a tensile force (pulling force) of 425 ft•lb (a ft•lb is the amount of energy needed to raise one pound up one foot).

a) Change 425 g to ounces.

b) Change 240 mm to inches.

c) If a .357 magnum shoots with a muzzle energy of 1,120 ft•lb, by how many times does its muzzle energy exceed the tensile strength of the handcuffs?

## 1.4 Unit Analysis and Certain Business Concepts

In this section we will combine unit analysis with certain concepts from business and industry. What results are some very useful tools. Tools that will serve you well whether you are a sergeant/supervisor on a police department, a restaurant manager, or an office manager. Please study them thoroughly.

Unit pricing is a concept that is used very heavily in business and industry. The term unit price means the price or cost of one unit of something. That unit might be just about anything... one liter, one kilogram, one gallon, or one piece. Examples 1 and 2 demonstrate how to find unit prices.

### Example 1
A woman bought 4.2 lb of beef for $12.39. What is the price per pound?

A unit price is the price for one unit of something. Here we want the price for 1 pound of beef. So the unit price is:

$$\frac{\$12.39}{4.2 \text{ lb}} = \frac{\$2.95}{\text{lb}}$$

### Example 2
A company bought 750 gallons of paint for $4,125.00. Find the unit price.

$$\frac{\$4,125.00}{750 \text{ gal}} = \frac{\$5.50}{\text{gal}}$$

Unit pricing can be used to determine which of two or more suppliers is offering the best price. Examples 3, 4, and 5 demonstrate this use of unit pricing.

### Example 3
A travel agency can order brochures from one printer in 500 copy lots for $245.00 and from another printer in 750 copy lots for $412.50. Which is the better buy?

Finding unit prices:

$$\frac{\$245.00}{500 \text{ copies}} = \frac{49¢}{\text{copy}} \quad \text{and} \quad \frac{\$412.50}{750 \text{ copies}} = \frac{55¢}{\text{copy}}$$

So the travel agency should buy the 500 copy lots as they are 6¢ per copy cheaper than the other lot.

## Example 4

A hotel manager can order bath towels in 1,000 piece lots from a Canadian supplier for $1,860.00 in Canadian dollars or from a US supplier for $1,660.00. If the currency exchange rate on that day was $1 Canadian = 82¢ US, which was the better buy? (This is without considering freight charges and any applicable taxes and/or tariffs.) If the manager marks-up the towels 150% for sale at the hotel's beach house, find the retail price of one towel.

Finding the unit prices:

$$\frac{\$1,860.00 \text{ (Can.)}}{1000 \text{ piece}} = \frac{\$1.86 \text{ (Can.)}}{\text{piece}} \quad \text{and} \quad \frac{\$1,660.00 \text{ (U.S.)}}{1000 \text{ piece}} = \frac{\$1.66 \text{ (U.S.)}}{\text{piece}}$$

Then changing the Canadian unit price to a U.S. unit price:

$$\frac{\$1.86 \text{ (Can.)}}{\text{piece}} \cdot \frac{\$0.82 \text{ (U.S.)}}{\$1.00 \text{ (Can.)}} = \frac{\$1.53 \text{ (U.S.)}}{\text{piece}}$$

Without freight, taxes or tariffs included it is clearly cheaper to buy from the Canadian supplier. Then marking up the towel 150% means the manager wants 100% of cost and a mark-up of 150% so he wants 250% of $1.53:

$$2.5 \cdot \frac{\$1.53}{\text{towel}} = \frac{\$3.83}{\text{towel}}$$

## Example 5

You can order beef from a supplier in quantities for $6.30 per kilogram or $2.95 per pound. Which is the better buy? By how many dollars per pound?

Since the second question specifies dollars per pound let's convert kilograms to pounds:

$$\frac{\$6.30}{1 \text{ kg}} \cdot \frac{1 \text{ kg}}{2.205 \text{ lb}} \approx \frac{\$2.86}{1 \text{ lb}}$$

Then $2.95/lb - $2.86/lb = $0.09/lb, so buying from the supplier who sells the beef for $6.30 per kilogram will save 9¢ per pound. For a restaurant that uses 20,000 pounds of this beef per year this represents an $1,800 annual savings.

The concept of unit pricing combined with the concepts of mark-up and mark-down or discount (from Pre-Algebra) gives us some powerful business analysis tools. Examples 6, 7, and 8 show you these very useful combinations.

## Example 6

Your motel buys candy in boxes which contain 50 one pound bags for $76.50. If your company directs that candy is marked-up 50%, find the retail price of one bag of candy.

Finding the price per one pound bag (unit price):

$$\frac{\$76.50}{50 \text{ lb}} = \frac{\$1.53}{\text{lb}}$$

Marking up the candy to find the retail price (marking up by 50% on cost means we want 100% of cost plus 50% so we want 150% of cost):

$$1.5 \bullet \frac{\$1.53}{\text{lb}} = \frac{\$2.30}{\text{lb}}$$

## Example 7

You are a restaurant manager. You buy cabernet wine in gallons for $13.20. You serve the wine in one liter carafes. If your company insists on a 60% mark-up (on cost) on wine, what is the retail price of the one liter carafe of cabernet?

Finding the unit price and converting:

$$\frac{\$13.20}{1 \text{ gal}} \bullet \frac{1 \text{ gal}}{3.781 \text{ L}} = \frac{\$3.49}{1 \text{ L}}$$

Then marking up by 60% on cost means we want 100% of cost plus 60%, so we want 160% of cost. So:

$$160\% \text{ of } \frac{\$3.49}{1 \text{ L}} = 1.6 \bullet \frac{\$3.49}{1 \text{ L}} = \frac{\$5.58}{1 \text{ L}}$$

## Example 8

You are a police sergeant. One of your duties is to supervise the annual firearms qualifications for all of the department's officers. You can buy ammunition from a local sporting goods store or from a wholesaler. The local store will sell you a 20 box case of .44 magnum cartridges that normally retails for $379.00 with a 15% discount, whereas the wholesaler will sell you the same case for $318.36 but also wants $22.00 shipping and handling. Which is the better buy?

Here we could do unit pricing on either one box or one case:

$$\frac{\$379}{\text{case}} \quad \text{vs} \quad \frac{\$318.36 + \$22.00}{\text{case}} = \frac{\$340.36}{\text{case}}$$

Then marking down the local store's price (a markdown of 15% means that you pay 100% - 15% or 85%):

$$0.85 \cdot \frac{\$379}{\text{case}} = \frac{\$322.15}{\text{case}}$$

So its actually cheaper to buy this case from the local store.

In Pre-Algebra we learned the following formulas:

**Profit or Loss = Selling Price - Cost**

$$\textbf{Percent Profit or Loss} = \frac{\textbf{Profit or Loss}}{\textbf{Cost}} \cdot \textbf{100\%}$$

Examples 9 and 10 demonstrate the use of these concepts with unit analysis.

## Example 9
A travel agency arranged a trip with transportation, meals, and lodging to Barbados for 13 couples for $60,000.00 cost to the agency. If the agency charged $3,200.00 per person, find the agency's profit per person and percent profit per person.

Finding the cost per person (unit cost):

$$\frac{\$60,000}{26 \text{ persons}} = \frac{\$2,307.69}{\text{person}}$$

Calculating the profit per person:

$$\frac{\$3,200}{\text{person}} - \frac{\$2,307.69}{\text{person}} = \frac{\$892.31}{\text{person}}$$

Finding the percent profit per person:

$$\frac{\dfrac{\$892.31}{\text{person}}}{\dfrac{\$2,307.69}{\text{person}}} \cdot 100\% = 38.7\%$$

## Example 10
A caterer served a wedding dinner for 70 people cost to the caterer of $360.00. If the caterer charged $6.50 per person, find the caterer's profit and percent profit.

Finding the caterer's cost per person:

$$\frac{\$360.00}{70 \text{ persons}} = \frac{\$5.14}{\text{person}}$$

Calculating the profit per person:

$$\frac{\$6.50}{\text{person}} - \frac{5.14}{\text{person}} = \frac{\$1.36}{\text{person}}$$

Finding the percent profit per person:

$$\frac{\dfrac{\$1.36}{\text{person}}}{\dfrac{\$5.14}{\text{person}}} \bullet 100\% = 26.5\%$$

# Exercises 1.4

Complete the following exercises. Round the result to the nearest cent if money or to the hundredths place if not money.

1.    A man bought 20 lb of nails for $15.60. What is the price per pound?

2.    A company bought 350 feet of chain link fence for $2,327.50. Find the price per foot.

3.    A hotel manager can order laundry detergent in 25 lb boxes for $31.12 or in 10 kg bags for $29.77. Which is the better buy?

4.    A restaurant manager can order goblets in 100 piece lots from a British supplier for £ 69 or from a U.S. supplier for $93.00. If the currency exchange rate on that day was $1.55 = £ 1, which was the better buy? (This is without considering freight charges and any applicable taxes and/or tariffs.)

5.    An office manager can order computer disks in 10 packs for $9.30 or in 25 packs for $22.25. Which is the better buy?

6.  A restaurant manager can order potatoes from a supplier in quantities for 70¢ per kilogram or 30¢ per pound. Which is the better buy?

7.  A travel agency can order discount amusement park tickets in 500 ticket lots for $990.00 or 2,000 ticket lots for $3,820.00. Which is the better buy?

8.  You are a restaurant manager. You buy white wine in gallons for $16.50. You serve the wine in one liter carafes. If your company insists on a 45% mark-up (on cost) on wine, what is the retail price of a one liter carafe of the white wine?

9.  You are an office manager. You can order computer disks in 10 packs for $9.30. An office next door wants to buy four disks from you. If you decide to mark-up the disks 25%, find the marked up price for four disks.

10. A travel agency arranges a tour bus trip to Nashville with meals and lodging for 45 people for $13,500 cost to the agency. If the travel agency charges 20% for their service, find the price of the trip for each person.

11. One unit of a chain store mismarked grass seed and sold it for $1.09 per kilogram rather than $1.09 per pound. If the grass seed costs 60¢ per pound and 1,500 kilograms were sold, how much money (on cost) was lost? What was the percentage loss?

12. A police chief reports that an average patrol officer drives 160 miles per shift. The city runs 21 shifts per week. Each shift has 6 cars. If the cars average 11 miles per gallon, how many gallons should the chief budget for the year? How many gallons should the chief request if he marks-up the budget by 10%?

13. A travel agency buys road maps in cartons of 500 for $275.00 with $14.00 shipping and handling. If they sell the maps for 95¢ each, find the agency's profit and percent profit on each map?

14. A caterer served a business dinner for 32 people at a total cost to the caterer of $112.00. If the caterer charged $6.00 per person, find the caterer's profit and percent profit.

15. A travel lodge manager buys hard candy in 20 kilogram boxes for $75.00. She then packages the candy in one pound bags which she sells for $3.99. Find her profit and percent profit.

16.    A restaurant buys bourbon in one liter bottles for $11.75 and sells one fluid ounce shots for $2.75 each. Find the profit per shot and percent profit per shot.

17.    You are a police lieutenant. One of your duties is to supervise the annual firearms qualifications for all of the department's officers. You can buy ammunition from a local sporting goods store or from a wholesaler. The local store will sell you a 20 box case of 12 guage shotgun shells that normally retails for $259.00 with a 12% discount, whereas the wholesaler will sell you the same case for $215.00, but also wants $18.00 shipping and handling. Which is the better buy?

18.    An upstate New York hotel manager can order bed spreads in 100 piece lots from a Montreal Canada supplier for $1,530.00 in Canadian dollars or from a US supplier for $1,290.00. If the currency exchange rate on that day was $1 Canadian = 83¢ US, which was the better buy? (This is without considering freight charges and any applicable taxes and/or tariffs.) A customer has asked the manager if she could buy one of the bed spreads. If the manager marks it up by 75% what is the price she charges the customer?

19.    The state allocated $31,050.00 per inmate to fund the department of corrections in 1997. If the state legislature provided a 3.5% increase for 1998, find the total amount allocated for 1998 if there are 39,250 inmates.

20. The state requires that there be one corrections officer per 36 inmates per shift. a) How many corrections officers are needed to staff a facility that has 360 inmates? (Remember there are three shifts per day and seven days per week and that each CO covers 5 shifts per week) b) What is the annual corrections officer payroll of the facility if the average pay of a corrections officer at that facility is $30,775.00? **NOTE: This is without considering vacations or absences.**

# 1.5 Review Exercises for Chapter 1

Write your answers in the spaces provided. If the item requires calculation show all your work in the space provided. Round answers to hundredths if rounding is needed. The answers may be found in the rear of the book.

## Vocabulary Exercises

1. Define each of the following terms:

    a) denominate number

    b) compound denominate number

    c) unit

    d) base unit

    e) auxiliary unit

    f) derived unit

    g) conversion

    h) reduction

## Compare and Contrast Exercise

2.    Compare and contrast the metric and customary systems of measurement. Your discussion should include base units, auxiliary units, and prefixes.

## Prefix Interpretation Exercises

Using your knowledge of the prefix values to answer each question.

3.    How many watts are contained in 100 kW?

4.    What fraction of a gram is 100,000 µg?

5.    How many ML are contained in 2,500 kL?

6.    How many dollars are contained in 25 Megadollars?

7.    What fraction of a liter is 100 mL?

8.    What fraction of a liter is 240 mL?

9.    What multiple of a volt is 0.050 kV?

10.    What multiple of a meter is 0.000005 Mm?

11.    How many meters in 2.2 hm?

12.    What fraction of an ampere is 200 mA?

13.    What multiple of a kilogram is 0.450 Mg?

14.    How many kilonewtons are contained in 3.5 MN?

## Unit Analysis Exercises

Do each of the following problems and round to the nearest cent if money and to the nearest hundredth if not money.

15.    Write 30.25 lb as a compound denominate number using pounds and ounces.

16.    Write 7.7 ft as a compound denominate number using feet and inches.

17.    Write 2.27 m as a compound denominate number using meters and centimeters.

18.    Write 16 lb 9 oz as a decimal denominate number with pounds as its only unit.

19.    Write 7 m 75 cm as a decimal denominate number with meters as its only unit.

20.    Write 50 yd 2 ft 3 in as a decimal denominate number with feet as its only unit.

21.    Convert 25 L to qt.

22.    Reduce 10,950 ft to mi.

23.    Convert 70 km to mi.

24.     Convert 90 floz to L.

25.     Convert 2.5 qt to cc.

26.     Reduce 27.4 cm to mm.

27.     Reduce 440 yd to mi.

28.     Reduce 2.25 days to hr.

29.     Reduce 16 hr to sec.

30.     Convert 48 mi to km.

31.     Reduce 50.4 square feet to square inches.

32.     Reduce 2,250 square centimeters to square millimeters.

33.     Convert 8.5 square centimeters to square inches.

34.     Convert 88.75 square meters to square feet.

35. Reduce 90 square yards to square feet.

36. A hotel is rectangular and measures 160 feet by 640 feet. How many acres does it cover?

37. Reduce 1.10 L to cubic centimeters.

38. Reduce 3,400 cubic centimeters to L.

39. Reduce 400 gal to cubic feet.

40. Convert 50 cubic meters to cubic feet.

41. Convert 240 cubic feet to cubic meters.

42. Convert 16 fluid ounces to cc.

43. Convert 250 mL to fluid ounces.

44. A box shaped shipping crate measures 5 ft by 3 ft by 2 ft. Find the volume in cubic inches.

45. A certain six cylinder engine has a displacement of 4.3 liters. Convert 4.3 L to cubic inches.

46. A swimming pool filtering system has a 100 gpm (gallons per minute) pump. If the pool is flat bottomed and measures 40 ft by 100 ft by 5 ft, how long will it take the pump to circulate all the water through the filter?

47. A hotel manager can order automatic dishwasher detergent in 100 liter drums for $810.00 or in 35 gallon drums for $685.05. Which is the better buy?

48. A travel agency can order a tour guide in 500 copy lots for $240.00 or in 800 copy lots for $768.00. Which is the better buy?

49. Population biologists often use an area density to describe populations. If a population biologist estimates there are 17 people per 12 acres in a certain region, how many people reside on 40 square miles within that region?

50. Convert 45 mi/hr to km/hr.

51. Convert the pressure 220 lb/in² to kg/cm².

52. A very fast bullet has a muzzle velocity of approximately 4,000 ft/sec. Reduce this speed to mi/hr.

53. The standard issue .357 caliber handgun issued by many police departments has a bullet muzzle velocity of approximately 1,100 ft/sec. Reduce this speed to mi/hr.

54. In the United States we measure automobile fuel efficiency in miles per gallon of gasoline consumed. Convert 32 mi/gal to km/L.

55. Which car is more fuel efficient...car A at 38 mi/gal or car B at 16 km/L? By how many mi/gal?

56. Convert the pressure 100 lb/in² to kg/m².

57. Convert the pressure 72.0 kg/m² to lb/in².

58.    A Xerox copier has a print rate of 48 pages per minute. How long will it take the copier to print 240 pages?

59.    A certain major metropolitan area reported a 1997 homicide rate of 112. Change 112 homicides per year to homicides per week.

60.    If the crime statistics for a certain city show 31 petty thefts per week, find the number of petty thefts per year.

61.    A salesperson for your company drove 450 kilometers one week. If she is to be reimbursed 20¢ per kilometer for her travel, how much should she be paid?

62.    If a 4 ounce portion of french fries contains 480 calories, how many calories are contained in 10 ounce portion of french fries?

63.    If a 4 ounce portion of ice cream has 500 calories, how many calories are contained in seven ounces of the ice cream?

64. If a 2.3 oz cereal bar has 230 calories, how many calories are contained in 6 ounces of the cereal bar?

65. In a certain city the police department reports that the average full-time patrol officer handles 3 petty theft calls per week. If the department has 14 full-time patrol officers, how many petty theft calls are handled per year?

66. A legal secretary and her new assistant can process 8 documents and 5 documents respectively each workday. How many days will it take them to process 65 documents?

67. A production line at a plastics plant has two molding machines that produce 300 and 240 pieces per hour respectively. How long will it take both presses working together to produce 2,700 pieces?

68. A stamping machine for making flat metal parts can produce 50 parts per minute. How many hours will it take the machine to produce 12,000 parts?

69. A motel contains 48 rooms which rent for $60.00 per day per room. If the average occupancy rate for the hotel is 55%, find the average monthly revenue for the motel.

70. A man bought 20 lb of pork chops for $34.00. What is the price per pound?

71. A woman bought 300 gallons of fuel oil for $336.00. Find the price per gallon.

72. A restaurant manager can order flour in 25 lb bags for $29.32 or in 10 kg bags for $25.47. Which is the better buy?

73. A restaurant manager can order plates in 50 piece lots from a German supplier for 360 D.M. (Deutsch Marks) or from a U.S. supplier for $175.00. If the currency exchange rate on that day was $1.00 = 2.4 D.M., which was the better buy? (This is without considering freight charges and any applicable taxes and/or tariffs.)

74. An office manager can order legal tablets in 10 packs for $11.00 or in 25 packs for $26.25. Which is the better buy?

75. A restaurant manager can order apples from a supplier in quantities for $1.50 per kilogram or 70¢ per pound. Which is the better buy?

76. A travel agency can order Rose Bowl tickets in 150 ticket lots for $3,150.00 or 250 ticket lots for $6,000.00. Which is the better buy?

77. You are a restaurant manager. You buy white wine in gallons for $16.50. You serve the wine in one liter carafes. If your company insists on a 45% mark-up (on cost) on wine, what is the retail price of one liter carafe of white wine?

78. A travel agency arranges a tour bus trip to Dollywood with meals and lodging for 35 people for $10,500 cost to the agency. If the travel agency charges 15% for their service, find the price of the trip for each person.

79. One unit of a grocery chain mismarked coffee beans and sold them for $1.99 per 3 lb can rather than $1.99 per pound. If the coffee costs 95¢ per pound and 220-3 lb cans were sold, how much money (on cost) was lost? What was the percentage loss?

80. A police chief reports that the average patrol officer drives 110 miles per shift. The city runs 21 shifts per week. Each shift has 5 cars. If the cars average 14 miles per gallon, how many gallons should the chief budget for the year? How many gallons should the chief request if he marks-up the budget by 15% to allow for emergencies?

81. A travel agency buys trip guides in cartons of 100 for $97.00 with $9.00 shipping and handling. If they sell the trip guides for $1.95 each, find the agency's profit and percent profit on each map.

82. A caterer served a luncheon for 25 people at a total cost to the caterer of $87.00. If the caterer charged $5.00 per person, find the caterer's profit and percent profit.

83. A travel lodge manager buys firewood in cords (128 cubic feet) for $80.00. She then ties firewood in 2 cubic foot bundles which she sells for $3.99. Find her profit and percent profit.

84. A restaurant buys wine in one liter bottles for $6.75 and sells 2 fluid ounce glasses of the wine for $2.00 each. Find the profit per glass and percent profit per glass.

85. If the state estimates that it costs $31,050.00 per inmate per year, find the cost of housing 635 inmates for five months.

86. The state requires that there be one corrections officer per 12 inmates per shift at their maximum security facilities. a) How many corrections officers are needed to staff a max facility that has 480 inmates? (Remember there are three shifts per day and seven days per week and that each CO covers five shifts per week) b) What is the corrections officer annual payroll of the facility if the average officer earns $31,250.00? c) What is the annual CO payroll of the facility if each officer gets a 5% raise? **NOTE: This is without considering vacations or absences.**

# CHAPTER 2
## INTRODUCTION TO ALGEBRA

INTRODUCTION:

*Algebra* is the branch of mathematics which generalizes the operations of arithmetic by allowing letters to take the place of numbers. For example, the statement:

$$2 + 3 = 3 + 2$$

can be generalized to:

$$a + b = b + a$$

for any two numbers a and b. We know the second statement as the commutative property of addition from basic math. Such generalizations allow us to convey a great amount of information in a very concise, condensed form.

OBJECTIVES:

Upon completion of this chapter you will be able to:

1. Explain what is meant by each of the terms: algebra, formula, general formula, specific formula, literal symbols, variable, coefficient, constant, algebraic term, algebraic expression, like terms, combining like terms, distributive rule, distribution, base, exponent, power, superscript, subscript, translation, square root, evaluation, and substitution.
2. Compare and contrast algebra and arithmetic.
3. Write formulas and evaluate formulas.
4. Recognize and combine like terms in algebraic expressions.
5. Use the distributive rule quickly and accurately.
6. Use the first two exponent rules quickly and accurately.
7. Use the distributive rule, the first two exponent laws, and combining like terms to simplify algebraic expressions.
8. Write algebraic expressions and evaluate algebraic expressions.

## 2.1 Formulas

The practice of using letters to stand in for numbers may have begun in an attempt to generalize results as is done with formulas. A *formula* is a concise, condensed set of mathematical instructions in equation form. For example, the statement that the area of a rectangle may be found by multiplying the length (L) times the width (W) is:

$$A = L \bullet W$$

There are two kinds of formulas, general formulas and specific formulas.
The formula for the area of a rectangle $A = L \bullet W$ is a *general formula* since it may be used to calculate the area of **any** rectangle.

For example, a rectangle with a length of 10 feet and a width of 6 feet has an area A = L•W = 10 feet•6 feet = 60 feet•feet = 60 ft$^2$.

To demonstrate what is meant by a specific formula, let's consider an example. Suppose a car rental company offers a special deal where a person may rent a car for $12 per day, 10 cents per mile, and a one-time service charge of $20.00. Then a formula for the cost (C) of the car rental is C = 12d + 0.10m + 20 where d is the number of days and m is the number of miles. This formula works only for this particular car rental deal and thus is called a *specific formula*.

In the formula C = 12d + 0.10m + 20, the letters C, d, and m are called *variables* (the cost C of the rental varies as both the number of days d and miles m driven vary) and the number 20 is called a *constant* because its value never varies. The numbers 12 and 0.10 are called coefficients. A *coefficient* tells us how many of a certain variable we have. For example, in the formula P = 3V + 5, we have three V's.

Formulas will play a large role in our studies and in our career work. We need to be able to write formulas, evaluate formulas, and algebraically manipulate formulas. We shall begin with formula writing and evaluation. We will take up algebraic manipulation of formulas in Chapter 3.

## Example 1
Write a formula for each statement. That is, translate each of the given statements into a formula.

a) Pressure is weight divided by area.
$$P = \frac{W}{A}$$

b) Power is work divided by time.
$$P = \frac{W}{t}$$

**Note:** It generally does not cause a problem when we allow the same letter to stand for two or more different quantities in different formulas as the correct formula choice will usually be obvious from the context of the problem.

c) The average speed of an object is the distance it travels divided by the time of travel.
$$s = \frac{d}{t}$$

d) A formula commonly used in business is part equals base times rate.

58

$$P = BR$$

e)  The potential energy an object has by virtue of its height above the earth's surface is the product of its mass, the acceleration (g) due to gravity, and the height of the object.

$$p = mgh$$

f)  The work done in lifting an object is the product of the object's weight and the height lifted.

$$W = wh$$

Here we need both the capital W and the lower case w. This example makes it easy to understand the following rule.

---

**Rule:**   Do not switch between capitals and lower case letters unless you are certain they represent the same quantity.

---

To evaluate a formula means we substitute some specified values into the formula and calculate the desired quantity.

## Example 2
Evaluate the given formula at the specified values.

a)  Evaluate the formula $F = 2v + 3r + 4$ if $v = 5$ and $r = 6$.

$$F = 2(5) + 3(6) + 4 = 10 + 18 + 4 = 32$$

b)  Evaluate the formula $T = ts - 3t + 4s - 3$ if $t = -1$ and $s = -4$.

$$T = (-1)(-4) - 3(-1) + 4(-4) - 3 = 4 + 3 + (-16) - 3 = -12$$

Formula evaluation problems will almost always be stated in a fashion similar to the following problems.

## Example 3
Calculate the required quantity.

a)  Calculate the pressure exerted on a floor by a 780 lb refrigerator standing on 4 circular leg tips that have a combined contact of 12 in² of floor area. Using $P = W/A$ from Example 1a:

$$P = \frac{780 \text{ lb}}{12 \text{ in}^2} = 65 \ \frac{\text{lb}}{\text{in}^2}$$

b) Calculate the average speed of a car that traveled 120 miles in 2.4 hours. Using s = d/t from example 1c:

$$s = \frac{d}{t} = \frac{120 \text{ mi}}{2.4 \text{ hr}} = 50 \frac{\text{mi}}{\text{hr}}$$

c) Calculate the cost of renting a car for six days from the car rental company in the introduction (Section 2.1) if the car was driven 1,842 miles.

Using C = 12d + 0.10m + 20 from the introduction we have:

C = 12.00•6 + 0.10•1,842 + 20.00 = $276.20

d) Calculate the work done by a clerk in lifting a 45 lb carton up 3 feet.

Using W = wh from example 1f:

W = wh = 45 lb • 3 ft = 135 ft•lb

Note: The ft•lb, pronounced foot pound, is the amout of work required to lift one pound up one foot.

e) The perimeter of a rectangle is given by P = 2L + 2W where L = length and W = width. Calculate the perimeter of a rectangle with length = 5.1 cm and width = 2.94 cm.

5.1 cm

2.94 cm

P = 2(5.1 cm) + 2(2.94 cm) = 16.08 cm which rounds to 16.1 cm (tenths)

f) Calculate the area of a rectangle with length 5.1 cm and width 2.94 cm. The area of a rectangle is given by A = l•w or lw where l = length and w = width.

A = 5.1 cm•2.94 cm = 15.0 cm² (to the nearest tenth)

Now that we have been introduced to formula writing and formula evaluation, let's take a look at some formulas that are used by people in many different careers.

## Example 4
A formula for calculating simple interest is $I = PRT$ where P equals the principal invested (or borrowed), R equals the interest rate (in decimal form), and T equals the time (in years).

   a)   Calculate the simple interest charged if you borrowed $7,500.00 at 9.9% per annum for 5 years.

   b)   What is the total payback?

   a) $I = 7,500.00 \bullet 0.099 \bullet 5 = \$3,712.50$

   b) Payback = Principal + Interest = $7,500.00 + $3,712.50 = $11,212.50

## Example 5
A formula for the retail price of any item is $R = C + MC$ where R = retail price, C = wholesale cost of the item and, M = the percentage mark-up (as a decimal). Calculate the retail price of an item if the cost is $4.25 and the mark-up is 60%.

$R = \$4.25 + 0.6 \bullet \$4.25 = \$4.25 + \$2.55 = \$6.80$

## Example 6
Consider the formula in Example 5. Sometimes the cost of a single item must be determined before we may use the formula $R = C + MC$. Suppose that UPS (United Parcel Service) just delivered a box that contains 50 Ohio maps. If the entire box of maps cost $17.50, UPS shipping cost $3.25, and your travel agency marks-up maps 40%, use the formula to calculate the retail price of a single map.

The total cost for all 50 maps is $17.50 + $3.25 = $20.75

Then the cost of a single map is $20.75/50 = 41.5¢ = 42¢

Using the formula for retail price we have:

$R = 42¢ + 0.4 \bullet 42¢ = 42¢ + 17¢ = 59¢$

## Example 7

One possible formula for depreciating the value of an item (a practice required by the federal government when writing off the cost of business equipment) is:

$$v = V(0.70)^t$$

where v = the depreciated value, V = the original value, and t the number of years.

Using the depreciation formula, calculate the value of a four year old piece of machinery that originally cost $7,200.00.

$$v = \$7,200.00(0.70)^4 = \$7,200.00(0.2401) = \$1,728.72$$

## Example 8

Police homicide detectives sometimes use the area of a circle to estimate the area covered by a blood stain. Once the size of the stain is determined, the detective can then consult a table and estimate the volume of blood lost by the victim.

The area of a circle is given by the general formula:

$$A = \pi r^2$$

Using the formula, calculate the area of a circular blood stain with a radius of 14 cm.

$$A = \pi(14 \text{ cm})^2 = \pi(196 \text{ cm}^2) = 615.8 \text{ cm}^2$$

## Example 9

A certain fast food restaurant chain has a formula that is used by their unit managers to order ground beef patties. It is:

$$M = \frac{T}{100} \bullet 5.5 \text{ lb}$$

where M = pounds of meat needed and T is the total sales for that order period. Use the formula to calculate the amount of ground beef needed for August if the prior August sales were $200,700.

$$M = \frac{200,700}{100} \bullet 5.5 \text{ lb} = 2,007 \bullet 5.5 \text{ lb} = 11,038.5 \text{ lb}$$

## Example 10

When business people determine the retail price of a good they must factor in their operating expenses. The operating expenses of a business consist of the fixed expenses such as cost of the mortgage and insurance and the variable expenses such as the cost of heat, light, water, and personnel. Suppose that a businesswoman determined that she must earn 30% on cost on all sales just to cover operating expenses. That is, if she sells all her goods at a 30% mark-up she will not make any profit but will merely **breakeven**.

If she wants to make a 20% profit on cost and operating expenses, calculate the retail price for an item that has a wholesale cost of $13.00.

First we must mark up the cost by thirty percent to cover operating expenses.
$$30\% \text{ of } \$13.00 \text{ is } 0.3 \bullet \$13.00 = \$3.90$$

Then to simply breakeven she needs cost plus operating expenses or:

$$\$13.00 + \$3.90 = \$16.90$$

Then marking up cost plus operating expenses by 20% we have:

$$20\% \text{ of } \$16.90 \text{ or } 0.2(\$16.90) = \$3.38$$

Then to pay the item cost, operating expenses, and make a 20% profit she needs:

$$\$16.90 + \$3.38 = \$20.28$$

Let's check the final percentage:

$$\frac{\$20.28 - \$13.00}{\$13.00} \bullet 100\% + 100\% = 156\% \text{ of cost.}$$

Can we write a formula that will do all the above calculations at once? The answer is yes.

First we must mark up the cost by thirty percent to cover operating expenses.

$$30\% \text{ of C is } 0.3C.$$

Then to simply breakeven she needs cost plus operating expenses or:

$$C + 0.3C \text{ or } 1.3C$$

Then marking up cost plus operating expenses by 20% we have:

$$20\% \text{ of } 1.3C \text{ or } 0.2(1.3C) = 0.26 \ C$$

Then to pay the item cost, operating expenses, and make a 20% profit she needs:

$$1.3C + 0.26C = 1.56C... \text{ So Retail price is } R = 1.56C$$

Let's use the formula to calculate the retail price.

$$R = 1.56(\$13.00) = \$20.28$$

Right now you may be asking this question. Why didn't we simply add the 30% mark-up and the 20% mark-up and mark-up each item by 50%? Well 50% of C is 0.5C and $C + 0.5C = 1.5 \ C$. But we have 1.56C. Where did the other 6% come from?

# Exercises 2.1

Please write your answers in the space provided. If the item requires calculation you are to show all work in the space provided. The answers may be found in the rear of the book.

In exercises 1 through 8 you are to identify the parts of each formula. Write coe for coefficient, con for constant, and var for variable.

1. $A = 2P + 3T + 1$   2. $R = 2rt - 4r + 7$   3. $F = -4g - 2t + 4$   4. $C = 4 - 3t$

5. $B = p + n - 9$   6. $D = 24 - T^2 + 6$   7. $t = 14 + 2s$   8. $Z = U^3$

In exercises 9 through 20 you are to translate the given statement into a **general formula** using the letters in parentheses.

9.  The area (A) of a triangle is one-half of the base (b) times the height (h).

10. The circumference (C) of a circle is given by twice the product of pi ($\pi$) and the radius (r).

11.  The perimeter (P) of a square is the product of (4) and the side length (S).

12.  The area (A) of a circle is found by multiplying pi ($\pi$) by the square of the radius (r).

13.  The surface area (A) of a sphere is given by four times the product of pi ($\pi$) and the square of the radius (r).

14.  The volume (V) of a right circular cylinder is the product of pi ($\pi$), the square of radius (r), and the height (h).

15.  The area (A) of rectangle is the product of the length (L) and the width (W).

16.  The volume (V) of a cone is one-third of the product of pi ($\pi$), the square of the radius (r), and the height (h).

17.  The perimeter (P) of a rectangle is twice the width (W) plus twice the length (L).

18.  The temperature in degrees Fahrenheit (F) is found by multiplying the temperature in degrees Celsius (C) by 1.8 and then adding 32 degrees.

19.  The perimeter (P) of a triangle is the sum of the side lengths (a), (b), and (c).

20.  The volume (V) of a box is the product of the length (l), the width (w), and the height (h).


In exercises 21 through 28 you are to translate the given statement into a **specific formula**. Please use letters that will help your reader follow your work. That is, use m for miles, p for price... letters which suggest what quantities they stand for.

21. If a certain taxi company charges a $2.00 flat fee plus 50 cents per mile, write a formula for the cost of a taxi ride.


22. An office machines maintenance company charges $30.00 for the service call plus $20.00 per hour for labor to service your office copier. Write a formula for the labor charge for a service call.


23. A plumber charges $60.00 for the first hour and $45.00 per hour after the first hour for labor on a night service call. Write a formula for her labor charge.


24. A computer disk manufacturer packs 10 disks in each box, 48 boxes in each carton, and 20 cartons on each pallet. Write a formula for the number of disks (N) in a shipment of n pallets.


25. A certain travel agent earns $6.00 per hour plus 0.5% commision on each transaction. Write a formula for her weekly pay. (Assume a 40 hour work week.)


26. The phone company charges your firm $26.00 per month plus 8¢ per call for local phone service. Write a formula for the cost of one month's local phone service.


27. An office machines rental company leases a top-of-the-line copier for $60.00 per month and 2¢ per copy. Write a formula for the cost of **one year's** rental.


28. A leasing company leases a cube van for $32.00 per day plus 28¢ per mile. Write a formula for the charge for leasing the van.

In exercises 29 through 59 you are to evaluate the given formula at the specified values.

29. Evaluate the formula B = 2c + 3d - 5    given c = 4 and d = 3.

30. Evaluate the formula T = 400t - 120r + 45    given t = 3 and r = 5.

31. Evaluate the formula P = 32t² + 5t + 12    given t = -2.

32. The area of a triangle is:
$$A = \frac{1}{2}bh$$
Find the area of a triangle with base = 7.78 in and height 5.8 in.

33. Using the formula of problem 32, find the area of a triangle with a base 12.275 in and height = 4.2 in.

34. Using the formula of problem 21, calculate the charge for a 12 mile taxi ride.

35. Using the formula of problem 22, calculate the cost of a 2.5 hour service call.

36. Using the formula of problem 23, calculate the cost of a 3 hour night service call.

37. Using the formula of problem 24, calculate how many disks are shipped in a shipment of 20 pallets.

38    Using the formula of problem 25, calculate the travel agent's pay if she made $32,060 in transactions in one 40 hour work week.

39.   Using the formula of problem 26, calculate the cost of two months local phone service if a total of 1,860 local calls were made.

40.   Using the formula of problem 27, calculate the cost of one year's rental if 60,400 copies were made in that year.

41.   Using the formula of problem 28, calculate the cost of a 3 day cube van rental if the van was driven 470 miles.

42.   A formula for converting the temperature in degrees Celsius to degrees Fahrenheit is:

$$F = \frac{9}{5}C + 32°$$

Find the Fahrenheit temperature if the Celsius temperature is 25 degrees C.

43.   Using the formula of problem 12, calculate the area of a circle with radius 2 feet.

44.   Using the formula of problem 9, calculate the area of a triangle with a base of 12 inches and a height of 4 inches.

45.   Using the formula of problem 17, calculate the perimeter of a rectangle that is 50 feet wide and 120 feet long.

46. Using the formula of Example 4, calculate the interest charged if you borrowed $4,000. 00 at 12% for 4 years.


47. Using the formula of Example 5, calculate the retail price of an item if its wholesale cost is $54.00 and the normal mark-up is 75%.


48. Using the formula of Example 4, calculate the payback for a loan if you borrowed $15,000 at 8.5% for 30 **weeks.**


49. Using the formula of Example 4, calculate the payback for a loan if you borrowed $11,000 at 6.75% for 18 **months.**


50. Using the formula of Example 7, find the depreciated value of a computer that cost $1,640.00 two years ago.


51. Your company just received a shipment of 25 laser printers. If the invoice for all 25 printers is $6,025.00, the shipping cost $63.00, and your company marks up the printers by 35%, use the formula of Example 6 to calculate the retail price of one printer.


52. Your restaurant just received a shipment of six-24 bottle cases of white wine. If the bill, including shipping, for the wine was $525.00 and your restaurant marks-up wine by 40%, use the formula of Example 6 to find the retail price of one bottle of the wine.

53. Using the formula of Example 7, find the depreciated value of an oven that had a cost of $995.00 three years ago.

54. Using the formula of Example 9, calculate the weight of beef patties to be ordered for December if the prior December sales were $316,450.

55. Study Example 10. Suppose a business owner has determined that her breakeven mark-up is 28%. Suppose that she wants to make a 30% profit on item cost and on operating expense. What retail price should she charge for an item that she buys from her wholesaler for $11.20?

56. Using the formula of Example 4, calculate the interest charged if you borrowed $45,000 at 8.9% for 145 **days**, use ordinary interest. (There are two ways to find interest by the day, ordinary uses 360 days in a year and exact uses 365 days in a year.)

57. Using the formula of Example 4, calculate the interest charged if you borrowed $6,800 at 10% for 234 **days**, use exact interest.

58. A department stores cost is $15 for a sweater. The original selling price is $45. At the end of the season, the store puts the sweaters on clearance at 75% off. Find the absolute loss. (absolute loss = cost – reduced selling.)

59. A stores cost of a coach is $599. To make room for a new model, the store marks the coach down 65%. Find the absolute loss if the original selling price was $899. (Use formula from problem 58.)

## 2.2 Algebraic Expressions and Combining Like Terms

In Section 2.1 we learned that algebra is generalized arithmetic which allows letters to stand in for undetermined numbers. We also learned that early attempts to generalize arithmetic were most likely motivated by a desire to pass on established results in formula form. Thus, formulas are the concise, polished result of much labor...a gift of knowledge from our predecessors.

Important as they are, formulas are not the primary focus of algebra– equation solving is. We begin our study of equation solving in Chapter 3. The remaining four sections of this chapter are intended to prepare us for our work in Chapter 3. Let's begin.

## Algebraic Terms

An *algebraic term* is any quantity made up of products, quotients, powers, and roots of numbers and/or literal symbols. Examples include:

$$3, \quad 9x, \quad 2x^2, \quad 3x^3y^4, \quad -2xy^3z^4, \quad 2\pi\sqrt{xy}, \text{ and } \quad 32x^3y^3z^3$$

## Algebraic Expressions

An *algebraic expression* is any combination of sums and differences of algebraic terms. All of the above examples are examples of one-term algebraic expressions. Examples of two-term algebraic expressions include:

$$9x - 5, \quad 2x^2 + 7, \quad 3x^2 - 23x, \quad x + 6, \quad -5y^3z + xz, \text{ and } \quad 32x^2y^2z^3 - 16\sqrt{xy}$$

Examples of three term expressions include:

$$3x^2 + 4x - 12, \quad 2x^2 - xy + 5y^2, \quad x^2 - y^3 + z^5$$

There is no restriction on the number of terms. For instance,

$$7x^5 - 4x^3 + 5x^2 + 5x - 3xy + y + 4y^3 - 3y^4 - 2y^2$$

is a nine-term algebraic expression.

The words variable, constant, and coefficient have the same meaning here with algebraic expressions as they did with formulas. For example, the algebraic expression:

$$3x^2 - 4x + 12$$

has variable x, constant 12, and coefficients 3 and -4.

## Like Terms

Like terms are those terms in an algebraic expression that differ only by coefficient.

## Example 1

    a)   2x, 3x, and 5x are like terms.

    b)   12x and $12x^2$ are not like terms.

    c)   $4x^2$, $-3x^2$, $\frac{1}{3}x^2$, and $\frac{x^2}{2}$ are like terms.

    d)   2xy and $2xy^2$ are not like terms.

    e)   2xy, 3kxy, and -35xy are like terms (where k is a constant).

    f)   $3x^3y$ and $4x^3y^2$ are not like terms.

    g)   $15x^3y^2$, $-7x^3y^2$, and $\frac{x^3y^2}{48}$ are like terms.

## Combining Like Terms

When we studied denominate numbers in Chapter 1, we learned that we may add or subtract only denominate numbers which have exactly the same unit. Adding or subtracting algebraic terms follows a similar rule...we may add or subtract only like terms.

## Example 2

    a)  As 3 ft + 2 ft = 5 ft, 3x + 2x = 5x.

    b)  As 7 ft2 + 9 ft2 = 16 ft2, $7x^2 + 9x^2 = 16x^2$.

    c)  We know that 3 ft and 5 ft2 can't be added because they are different physical quantities. Similarly, 3x and $5x^2$ can't be added because they are not like terms.

    d)  We know that 10 ft and 24 ft•lb can't be added because they are different physical quantities. Similarly, 10x and 24xy can't be added because they are not like terms.

**Example 3**
Combine like terms where possible.

a)  $2x + 4y + x + 5y$
    $= 3x + 9y$

b)  $4x - 2y + 7x + 8y$
    $= 11x + 6y$

c)  $12x^2 + 7x - 6x^2 - 3x$
    $= 6x^2 + 4x$

d)  $2a - 3c + 4b + 6a + 25b$
    $= 8a + 29b - 3c$

e)  $2a^2 - 3b + 2 - 3a^2 + 7b - 5$
    $= -a^2 + 4b - 3$

f)  $3x^2 + 4x + 6 - x^2 - 5x + 12$
    $= 2x^2 - x + 18$

g)  $2a^2 - ax + x^2 + 4ax + a^2$
    $= 3a^2 + 3ax + x^2$

We often refer to algebraic expressions by the number of terms. According to this scheme, $2x - 5$ is a 2 term expression, $5x^2 - 3x + 7$ is a 3 term expression, and $4x^2 - 3xy + 2x - 4y^2$ is a 4 term expression. We must make sure that all like terms have been combined before counting the number of terms. For example, $3x^2 + 5x + 2x^2 - 3x + 1$ is not a 5 term expression, but is actually $5x^2 + 2x + 1$, a 3 term algebraic expression.

# Exercises 2.2

In exercises 1 through 39 you are to combine like terms where possible. Show all work in the spaces provided. The answers may be found in the rear of the book.

1.  $3x + 2 + 5x + 6$

2. $4y - 7 + 3y + 5$

3. $12t - 5q + 3t + 8q$

4. $-5p - 3r + 7p + 2r$

5. $3x + 5 - 7x - 12$

6. $4x^2 - x + x - 4x^2$

7. $6x - 3 - 7x + 6$

8. $\frac{1}{2}x + 3 + \frac{5}{2}x - 2$

9. $7x^2 + 3x - 5x^2 + 2x$

10. $2xy - x + 6xy - 6x$

11. $2x + 3y - 4x + 5y + 3x - 7y$

12. $a + 2b - 3c + d - 4c + a - b + 6d$

13. $6x^2 + 4x - 3 + 2x^2 - 7x + 12$

14. $16x^2 + 2xy - 3x^2 - 9xy + y^2$

15. $12x^2 + 3x - 5x^2 - 7x + 6$

16. $11x^2 + x - 3xy - y + 3x + 4xy + 6y - 5x^2$

17. $32a^2 + 14a - 5 + 13a^2 - 8a + 12$

18. $19t^3 - 4r^2 + 3t^3 + 11r^2$

19. $4s^3 + 3s^2 - 4s + 7 - 2s^2 + 3S^3 + s - 9$

20. $12p^2 - 2p - 7 - 5p^2 - 7p + 3 - p^2 - 11$

21. $4xy^2 + 5xy + 7xy^2 + 2xy + y^2 - 7y$

22. $pv + pv^2 - p^2 + v^2 + 6pv - 3v^2$

23. $kAT - 4T + 5kAT - 3A + 4kAT - 11T$

24. $mv^3 - mv^2 + mv - 2mv^2 + 3mv^3 + 11mv$

25. $R^2 - r^2 + 3R - 7r + 4R^2 - 9r^2 + R - 13r$

26. $7t^2 - 3t + 4 + 4t^2 + 3t - 5 + 3t^2 - 4t + 6 + 3t - 2$

27. $6q^2 + 11q - 5 + 32q^2 - 21q + 14q^2 - 23q + 4 - 17q$

28.  $0.6r^2 - 0.05r + 1.2 + 2.2r^2 + 0.25r - 0.8$

29.  $0.9s^2 + 1.06s - 0.04 + 1.3s - 0.4s^2 + 1.6$

30.  $\dfrac{1}{2}x^2 - \dfrac{3}{4}x + \dfrac{5}{2}x^2 - \dfrac{5}{4}x$

31.  $4p^2 + 0.75p - 3.25p^2 + 2.25p - 0.75$

32.  $\dfrac{7}{8}r^2 - \dfrac{5}{6}r + \dfrac{17}{8}r^2 - \dfrac{7}{6}r$

33.  $2a^2cp - 3acp + 4a^2cp - 3cp + 2acp + 2cp - 3c$

34.  $12b^3c^2 - 3b^2c^2 + 4bc^2 - 9b^3c^2 - 5bc^2 + 11b^2c^2$

35. $3xy - 2xy^2 + 4x^2y + 7xy - 4x^2y + 11xy^2$

36. $4abc - 2c^2 + 3ac^2 - 3abc + 4c^2 - 6ac^2$

37. $\dfrac{3}{5}s^2 + \dfrac{2}{3}s + \dfrac{12}{5}s^2 - \dfrac{5}{3}s + 6$

38. $11x + 22y + 33z - 6x - 17y - 12z - 5x - 5y - 21z$

39. $9^2y^2 - 3^2y + 4^2y^2 + 7^2y - 31y^2 + 24y + 11$

## 2.3 The Distributive Rule

When we studied the order of operations in earlier courses, we learned that the grouping symbols

$$( \; ), \; [ \; ], \text{ and } \{ \; \}$$

are used to indicate that we should do all the calculations within the grouping symbols first. For example:

$$2(3 + 4) = 2(7) = 14$$

But there is another way to do this problem:

$$2(3 + 4) = 2(3) + 2(4) = 6 + 8 = 14$$

While this is a violation of operations order, it does work. This method of calculating 2(3 + 4) is an example of an algebraic rule called the distributive rule.

## The Distributive Rule

If a, b, and c are any three numbers, then

$$a \, (b \pm c) = ab \pm ac$$

We call this the distributive rule because the factor outside of the grouping symbols is distributed to each of the terms within the grouping symbols. Since we always follow operations order, we rarely use the distributive rule on expressions that contain only numbers. The value of the distributive rule is that it allows us to eliminate grouping symbols when we are unable to evaluate the expression within the grouping symbols. For example, in the expression 3(x + 5) we can't add x and 5 since we don't know the value of x, but we can distribute the 3 and obtain the expression 3x + 15, which is free of parentheses. Eliminating grouping symbols allows us to combine like terms with terms outside the grouping symbols.

## Example 1
Distribute where needed and then combine all like terms.

    a)  3(x + 4) + 2x - 8
         = 3x + 12 + 2x - 8
         = 5x + 4

    b)  4(2x + 3) + 6 - 5x
         = 8x + 12 + 6 - 5x
         = 3x + 18

    c)  -3(5x - 4) + 3x - 16

$$= -15x + 12 + 3x - 16$$
$$= -12x - 4$$

d)   $-10(2x^2 - 4x) + 25x^2 - 36x$
$$= -20x^2 + 40x + 25x^2 - 36x$$
$$= 5x^2 + 4x$$

There is no limitation on the number of terms within the grouping symbols. There might be ten terms or twenty terms inside the grouping symbols...the number doesn't matter as long as we distribute the factor outside the grouping symbols to each term within the grouping symbols.

## Example 2
Simplify as completely as possible. This means distribute and then combine like terms.

a)   $3x^2 + 14x + 6 + 2(4x^2 - 2x + 6)$
$$= 3x^2 + 14x + 6 + 8x^2 - 4x + 12$$
$$= 11x^2 + 10x + 18$$

b)   $4(x^2 - 5x - 2) - 3x^2 + 19x - 7$
$$= 4x^2 - 20x - 8 - 3x^2 + 19x - 7$$
$$= x^2 - x - 15$$

There are no limitations on either the number of variables or the number of distributions in a problem.

## Example 3
Simplify as completely as possible.

a)   $3b - 3c + 2(a^2 + b - 2c)$
$$= 3b - 3c + 2a^2 + 2b - 4c$$
$$= 2a^2 + 5b - 7c$$

b)   $3(4x^2 - 7x - 5) - 2(2x^2 + 3x - 5) - 4x^2 + 6x + 24$
$$= 12x^2 - 21x - 15 - 4x^2 - 6x + 10 - 4x^2 + 6x + 24$$
$$= 4x^2 - 21x + 19$$

c)    $-2(a^2 - 3b + 2c - 5d) + 5a^2 + 8b + 5c - 11d$

$$= -2a^2 + 6b - 4c + 10d + 5a^2 + 8b + 5c - 11d$$

$$= 3a^2 + 14b + c - d$$

## Reminder

Exponents don't change when we combine like terms.

d)    $3(4x^2 - 3y^2 + 2z^2) + 22x^2 - 15y^2 + 13z^2$

$$= 12x^2 - 9y^2 + 6z^2 + 22x^2 - 15y^2 + 13z^2$$

$$= 34x^2 - 24y^2 + 19z^2$$

e)    $-5(2x^2 - 3x) - 2(5x^2 - 2x) - 3(2x^2 - 5x)$

$$= -10x^2 + 15x - 10x^2 + 4x - 6x^2 + 15x$$

$$= -26x^2 + 34x$$

A distribution that often confuses students is the distribution of -1.

## Example 4
Simplify as completely as possible.

a)    $(-1)(2x - 3)$
       $= -2x + 3$

b)    $-(2x - 3)$
       $= -2x + 3$

c)    $1 - (2x - 3)$
       $= 1 - 2x + 3$
       $= 4 - 2x$

d)    $7x + 8 - (3x - 5)$
       $= 7x + 8 - 3x + 5$
       $= 4x + 13$

e)    $4x^2 - 3x + 5 - (2x^2 - 8x - 7)$
       $= 4x^2 - 3x + 5 - 2x^2 + 8x + 7$
       $= 2x^2 + 5x + 12$

Examples 4c, 4d, and 4e show us how to subtract one algebraic expression from another. Example 5 will demonstrate more of these subtractions.

**Example 5**
Perform the stated operations.

a) Subtract the quantity 3x - 2 from the quantity 5x + 4.
5x + 4 - (3x - 2)
= 5x + 4 - 3x + 2
= 2x + 6

b) Subtract three times the quantity x2 + 4x + 5 from five times the quantity 2x2 - 3x - 2.

$$5(2x^2 - 3x - 2) - 3(x^2 + 4x + 5)$$
$$= 10x^2 - 15x - 10 - 3x^2 - 12x - 15$$
$$= 7x^2 - 27x - 25$$

# Exercises 2.3

In exercises 1 through 35 you are to simplify as completely as possible. Show all work in the spaces provided. The answers may be found in the rear of the book.

1. 2x - 3 + 5(x + 2)

2. 3(x +4) + 5x - 6

3. 2(4x - 5) + 3(6 - x)

4. -4(2x - 3) + 2x - 1

5. -5(4 - 5x) + 3x + 12

6. $-2(12x - 5) - 3(4x - 5)$

7. $3(2x^2 + 4x + 3) - 4x - 5$

8. $7x^2 - 3(x^2 + 4x + 3) - 5x + 7$

9. $12x^2 - 3x - 4(2x^2 + 3x - 5) + 2x - 4$

10. $2a + 3b - c + 4(a - 2b + 3c)$

11. $3(x - 3) + 2(3 - x) + x + 3$

12. $4s - 3t + 2(3s - t) + 3(t + 3s)$

13. $-(3x - 2)$

14. $2 - (3x - 2)$

15. $x + 3 - (x + 3)$

16. $x + 3 - (x - 3)$

17. $4x - 7 - (3x - 7)$

18. $3x - (x - 3) + 2x - 1$

19. $4x - 1 - (2x - 1) + x - 7$

20. $-(3x - 1) - (2x - 1) - (2 - 5x)$

21. $2x^2 - 3x - 5 - (x^2 - 5x - 7)$

22. $3x^2 - 4x + 5 - (-2x^2 + 7x - 3)$

23. $7x^2 - 4xy + y^2 - (3x^2 - 5xy - 2y^2)$

24. $a^2 - 3b + c - (6c - 2b - 10a^2)$

25. $r^2s^2 - 3rs + s^2 - (r^2s^2 - 7rs - 2s^2)$

26. $1.6x - 2.4y - (1.4x + 0.8y)$

27. $2.2x - 3.4 - (0.7x - 7.8)$

28. $2.4(0.5x - 0.1y) - (2.2x - 3.8y)$

29. $1.5(3.2x + 2.3y) - 2(4.1x - 3.2y)$

30. $13m - 2.4n - 3(3.2m - 4.1n)$

31. $6x^2 - 4.5x - 3.2 - 4(0.75x^2 + 2.2x - 1.8)$

32. $-(4p^2 - 3) - 2(-2p^2 + 6) + 3$

33. $3a^2 - b^2 - 4c^3 - 3(2a^2 - 5b^2 - 2c^3 - 5)$

34. $2(y - w) - 2(3y - w) - (-w - 2y)$

35. $0.2x - 0.3y - (0.4x - 0.3y) - 2(0.5x - 0.3y)$

In exercises 36 through 51 you are to perform the stated operations.

36. Subtract the quantity $2x + 5$ from the quantity $4x - 7$.

37. Subtract the quantity $4x - 3$ from the quantity $7x - 8$.

38. Subtract the quantity $3x^2 - 4x - 7$ from the quantity $5x^2 - 3x + 12$.

39. Subtract the quantity $7x^2 - 5x - 3$ from the quantity $8x^2 - 15x + 6$.

40. Add twice the quantity $3x + 5$ to the quantity $4x - 7$.

41. Add three times the quantity $6x + 1$ to the quantity $11x - 2$.

42. Subtract four times the quantity $2x + 3$ from twice the quantity $6x - 11$.

43. Subtract three times the quantity $2x - 5$ from five times the quantity $4x + 7$.

44. Subtract twice the quantity $2x^2 - 4x - 3$ from three times the quantity $5x^2 + 7x - 5$.

45. Subtract four times the quantity $3x^2 - 5x - 3$ from seven times the quantity $2x^2 + 4x - 5$.

46. Multiply the quantity $3x^2 - 5x - 7$ by $(-1)$ and add that product to the quantity $14x^2 - 11x + 24$.

47. Multiply the quantity $x^2 - 3xz + z^2$ by $(-2)$ and add that product to the quantity $4x^2 - 5xz + 7z^2$.

48. Subtract the quantity $12x^2 - 4x + 3$ from the sum of the quantities $6x^2 + 5x - 3$ and $7x^2 + 6x + 12$.

49. Subtract the quantity $10x^2 - 2x + 5$ from the sum of the quantities $2x^2 + 5x + 12$ and $5x^2 + 3x + 18$.

50. Subtract four times the quantity $3x - 5$ from five times the sum of the quantities $2x + 3$ and $7x + 5$.

51. Subtract three times the quantity $2x + 3$ from five times the sum of the quantities $5x - 3$ and $2x - 5$.

## 2.4 More about the Distributive Rule

In the previous section we learned about a very important algebraic rule known as the distributive rule. This rule allowed us to eliminate grouping symbols so that we could combine like terms. For instance,

$$3(2x + 3) - 4(x - 5) - (5x - 7) + 2x + 1$$
$$= 6x + 9 - 4x + 20 - 5x + 7 + 2x + 1$$
$$= -x + 37$$

Like the example above, all of our work in Section 2.3 consisted of distributing only constants. Now we shall begin work that will require us to distribute variables. This work is easy with the help of the first exponent rule.

## Exponent Rule 1

If a is any number and m and n are positive integers, then:
$$a^m \bullet a^n = a^{m+n}$$
Here a is called the base and m and n are called exponents or powers.

## Example 1

a)  $2^3 \bullet 2^4 = 2^7$ Let's prove this is so.

   $2^3 \bullet 2^4 = (2 \bullet 2 \bullet 2)(2 \bullet 2 \bullet 2 \bullet 2) = 2 \bullet 2 \bullet 2 \bullet 2 \bullet 2 \bullet 2 \bullet 2 = 2^7$

b)  $3^4 \bullet 3^5 = 3^9$

c)  $2^2 \bullet 3^2 \neq 6^4$ These terms do not have the same base, so the rule
   does not apply.

d)  $2x^3 \bullet x^4 = 2x^7$

e)  $3x^2 \bullet 4x^5 = 3 \bullet 4 \bullet x^2 \bullet x^5 = 12x^7$

f)  $(2x^2)(5x^4)(-5x^2)(3x^5) = -150x^{13}$

Examples d, e, and f show that we multiply the coefficents and add the exponents when we multiply terms.

Let's use the first exponent law to help us with distribution.

## Example 2
Perform each distribution.

a) $x(2x + 3)$    Note: $x = x^1$

$= 2x^2 + 3x$

b) $2x(3x^2 - 6x + 4)$

$= 6x^3 - 12x^2 + 8x$

c) $-4x(x^2 - 5x - 3)$

$= -4x^3 + 20x^2 + 12x$

Let's tackle some problems that will require us to do both distribution and combining like terms.

## Example 3
Simplify as completely as possible.

a) $2x(3x + 4) + 2x^2 - 7x - 5$

$= 6x^2 + 8x + 2x^2 - 7x - 5$

$= 8x^2 + x - 5$

b) $-3x(4x - 5) - x(2 - x) + 3(2x^2 - 5x) - 5x + 3x^2$

$= -12x^2 + 15x - 2x + x^2 + 6x^2 - 15x - 5x + 3x^2$

$= -2x^2 - 7x$

Many applications require more than one variable. Such problems are easily handled variable by variable using the first exponent rule.

## Example 4
Simplify as completely as possible.

a)    $3x^2y^2 \bullet 4x^3y = 12x^5y^3$

b)    $-2x^2y(-3x^5y^2) = -2 \bullet -3 \bullet x^2 \bullet x^5 \bullet y \bullet y^2 = 6x^7y^3$

c)    $(-2r^2s^3t^5)(-5r^4s^2t^3) = 10r^6s^5t^8$

90

d) $-2xy(-5x^2y)(-3x^3y^5)(y^2) = -30x^6y^9$

e) $-3(4pq)(-2p^2q^5) = 24p^3q^6$

Let's tackle some distribution problems that contain multiple variables.

## Example 5
Simplify as completely as possible.

a) $3x(x+y) + 4x^2 - 2xy$
   $= 3x^2 + 3xy + 4x^2 - 2xy$
   $= 7x^2 + xy$

b) $4x(x-3) + 2x^2 + 4x + 5$
   $= 4x^2 - 12x + 2x^2 + 4x + 5$
   $= 6x^2 - 8x + 5$

c) $-3s(2s-4) + 6s^2 - 7s + 3$
   $= -6s^2 + 12s + 6s^2 - 7s + 3$
   $= 5s + 3$

d) $4t(t^2 - 3t + 4) + 3t^2 - 5t + 6$
   $= 4t^3 - 12t^2 + 16t + 3t^2 - 5t + 6$
   $= 4t^3 - 9t^2 + 11t + 6$

Some problems we encounter may require the use of the second exponent law.

## Exponent Rule 2

Let a be any number other than zero and let m and n be any positive integers. Then:

$$\frac{a^m}{a^n} = a^{m-n}$$

## Example 6
Simplify as completely as possible:

a) $\dfrac{2^5}{2^3} = 2^{5-3} = 2^2 = 4$

b) $\dfrac{x^7}{x^3} = x^4$

c) $\dfrac{-35x^9}{14x^4} = \dfrac{-5x^5}{2}$

d) $\dfrac{(2x-3)^4}{(2x-3)^3} = (2x-3)^1 = 2x - 3$

e) $\dfrac{x^4}{x^4} = x^{4-4} = x^0 = 1$

Example 6e demonstrates an interesting consequence of the second exponent rule.

We know that any number (except 0 ) divided by itself is 1. Therefore, since $\dfrac{x^4}{x^4} = 1$, it must be true that $x^0 = 1$.

## Exponent Rule 3

If a is any number other than zero, then:
$$a^0 = 1$$

Try this with your calculator:
enter 4...press $y^x$ ( or $xy$)... enter 0...press =

You have just calculated $4^0 = 1$. Now try this with your calculator:
enter 0...press $y^x$...enter 0... press =

You have just attempted to calculate $0^0$. The calculator's display should show the letter E or the word Error.

Example 6d of the preceding example shows us that the second exponent law applies even if the base is itself an algebraic expression. We will see more such problems in example 7.

## Example 7
Simplify as completely as possible.

a) $\dfrac{ax^3}{x^2} + a(x-2) + 4a$

$= ax + ax - 2a + 4a$

$= 2ax + 2a$

b) $12(4x^2 - 5x) - \dfrac{48x^4}{6x^2} + \dfrac{7a^2x^3}{a^2x^2}$

$= 48x^2 - 60x - 8x^2 + 7x$

$= 40x^2 - 53x$

c) $\dfrac{a^2}{a^2} - \dfrac{33ax^3}{11ax^2} + \dfrac{2a^2}{a^2}(x-2)$

$= 1 - 3x + 2(x - 2)$

$= 1 - 3x + 2x - 4$

$= -x - 3$

d) $\dfrac{25a(3x-y)^4}{5a(3x-y)^3} - \dfrac{bx^3}{bx^2} + \dfrac{12c^2}{-3c^2}(x+2y)$

$5(3x - y) - x - 4(x + 2y)$

$15x - 5y - x - 4x - 8y$

$10x - 13y$

e) $\dfrac{36x(x+2)^3}{12x(x+2)^2} - \dfrac{16a^4x^5}{8a^4x^4} - \dfrac{32x^5(x-1)}{16x^5}$

$= 3(x + 2) - 2x - 2(x - 1)$

$= 3x + 6 - 2x - 2x + 2$

$= -x + 8$

It is often convenient to use subscripts on one variable rather than using different variables. For example, a series of ten temperature measurements may be:

$$T_1, T_2, T_3, ..., T_{10}$$

rather than:
$$a, b, c, ..., j.$$

This practice allows us to use a single letter that quickly reminds us of what quantity it represents...in this case temperature.

Subscripts are never used in calculations. Where the 3 (a superscript) in $T^3$ means T to the third power, the 3 in $T_3$ is only there to distinguish $T_3$ from other subscripted T's.

### Example 8
Simplify as completely as possible.

a) $2(T_2 + T_3) - T_2 + T_3$
$$= 2T_2 + 2T_3 - T_2 + T_3$$
$$= T_2 + 3T_3$$

b) $V_1(V_1 + V_2 + k) + 2V_1^2 - 3kV_1$
$$= V_1^2 + V_1V_2 + kV_1 + 2V_1^2 - 3kV_1$$
$$= 3V_1^2 + V_1V_2 - 2kV_1$$

c) $4cs_1(2s_2 - s_1^3t) + 18cs_1s_2 - 6cs_1^4t$
$$= 8cs_1s_2 - 4cs_1^4t + 18cs_1s_2 - 6cs_1^4t$$
$$= 26cs_1s_2 - 10cs_1^4t$$

# Exercises 2.4

In exercises 1 through 46 you are to simplify as completely as possible. Show all work in the spaces provided. The answers may be found in the rear of the book.

1.  $x(3x - 5)$

2.  $2x(4x + 7)$

3.  $3x(5x - 2)$

4.  $5x(7x - 3)$

5.  $2x(x^2 - 3x - 5)$

6.  $3x(4x^2 + 5x - 7)$

7.  $4x(3x^2 - 11x - 2)$

8.  $-5x(7x^2 - 3x + 16)$

9.  $-2x(x^2 + 3x - 5) + 3x^2 + 6x - 7$

10.  $2x^2(x - 1) + 3x(x^2 + 4x) + 6$

11.  $12x(1 - x) + 3x(7 + x) - 10x^2 + 6x$

12.  $4x(x^2 + x - 3) + 2x(4 - x - x^2)$

13.  $(-3x^2)(-4x^5)$

14.  $(2x^5)(7x^3)$

15. $(-32x^3)(-x^2)(-2x^4)$

16. $-7x^2(-3x^3)$

17. $(3m^4n^2)(-5m^2n^3)$

18. $(-3m)(-4mn^2)(-n^4)$

19. $(-2mn^2)(-3m^2n)(-m^5)$

20. $4x^2(2x-y)+5x^3-7x^2y$

21. $3r(5r+t)+2t(r-2)+4r^2$

22. $a^2(2a^2+2b-3c)+2c(a^2-5b)+5a^2c$

23. $-v(v^2+u)-uv+3v^2$

24. $2x^2(x+xy-y)+4x^3+7x^2y+5x^3y$

25. $\dfrac{x^4}{x^3}$

26. $\dfrac{64y}{-16}$

27. $\dfrac{8x^5}{4x^2}$

28. $\dfrac{28t^7}{-4t^4}$

29. $\dfrac{42x^5}{-7x^3}$

30. $\dfrac{105x^3}{21x}$

31. $\dfrac{45x^3y^5}{-9x^2y^3}$

32. $\dfrac{32x^{11}y^9}{-2x^9y^4}$

33. $\dfrac{33x^5}{11x^4} + 2x$

34. $\dfrac{3x^2y}{x^2} + \dfrac{4y^2}{2y}$

35. $\dfrac{21x^3y^2}{7x^3y} + 2(3y - 5)$

36. $\dfrac{32a^2bc}{-8a^2c} - 4c(2 - c)$

37. $\dfrac{b^2}{b^2} - \dfrac{3a^2x^2}{ax} + a(2x) + 3$

38. $\dfrac{12t(x - y)}{4t} + \dfrac{16s^2(x + y)}{8s^2}$

39. $\dfrac{a}{a} + \dfrac{2b^3}{b^3} - \dfrac{4k^2x}{2k^2x} + \dfrac{b^3}{b^3}$

40. $\dfrac{bx^3}{bx} - 3x(4x - 5) + 2x^2 - 5x$

41. $\dfrac{q^2y^4}{q^2y^2} - 7y^2 - 3y(2 - 3y) - 4y$

42. $\dfrac{18(a - b)^2}{9(a - b)} + 2a - 3b$

43. $\dfrac{36(r-2s)^3}{-9(r-2s)^2} - \dfrac{2r^2s}{rs} + \dfrac{6rs^2}{3rs}$

44. $3T_1^2 + T_1 - T_1^2 + 5T_1$

45. $4R_1(2R_1 - 3R_2) + 2R_1^2 - 5R_1R_2$

46. $5s_1^2(3s_1^3 - 5s_1^2 + 4s_1) - s_1^5 + 3s_1^3$

## 2.5 More about Algebraic Expressions

In Section 2.1 we learned how to write formulas and how to evaluate formulas when the values of the variables were specified. In this section we will learn how to write algebraic expressions and how to evaluate algebraic expressions when the values of the variables are specified.

### Writing Algebraic Expressions

There are many different ways to describe mathematical operations. Perhaps the best method for learning these descriptions as well as the differences among them is to study many examples. Let's begin.

### Example 1
Translate each of the given statements into an algebraic expression. Let n be the number.

a)  Three more than the number
   $n + 3$ or $3 + n$

b)  Twice the number
   $2n$ (not $n2$ or $n \bullet 2$)

c)  One half of the number
   $\frac{1}{2}n$ or $\frac{n}{2}$

d)  Five less than the number
   $n - 5$ (not $5 - n$)

e)  Five less the number
   $5 - n$ (not $n - 5$)

f)  The number increased by 3
   $n + 3$ or $3 + n$

g)  The number decreased by 5
   $n - 5$ (not $5 - n$)

h)  Five more than three times the number
   $3n + 5$ or $5 + 3n$

i)  Four less than six times the number

6n - 4 (not 4 - 6n)

j)   One fifth the number decreased by three

$$\frac{1}{5}n - 3 \text{ or } \frac{n}{5} - 3 \text{ ... not } 3 - \frac{1}{5}n \text{ nor } 3 - \frac{n}{5}$$

k)   Five less one third the number

$$5 - \frac{1}{3}n \text{ or } 5 - \frac{n}{3} \text{ ...not } \frac{1}{3}n - 5 \text{ nor } \frac{n}{3} - 5$$

The above examples should remind us of the importance of commutativity. Remember that addition is commutative but subtraction is not. That is why n + 3 = 3 + n, but n - 5 ≠ 5 - n.

The words sum and difference don't simply mean addition and subtraction but mean the result of each operation. For example, in the statement 3 + 2 = 5, 3 + 2 is called the implied sum, and 5 is the sum. Since the terms sum and difference mean the result of addition and subtraction respectively, we must use grouping symbols whenever we encounter these terms.

## Example 2
Translate each of the given statements into an algebraic expression. Let n = the number.

a)   Three times the sum of the number and four

3(n + 4) or (n + 4)3
(the second is not as desirable as the first)

After distributing, 3(n + 4) = 3n + 12. This is clearly different from the statement: Three times the number increased by four or 3n + 4.

b)   Five times the difference of the number and three

5(n - 3) or (n - 3)5

After distributing 5(n - 3) = 5n - 15, it is easy to see that this new expression is very different from the statement: Five times the number decreased by three or 5n - 3.

The product and quotient, the results of multiplication and division respectively, do not require the use of grouping symbols since normal operations order specifies that multiplication and division come before addition and subtraction.

101

## Example 3
Translate each of the given statements into an algebraic expression. Let n = the number.

a) Four more than the product of six and the number

$$(6n) + 4 \text{ or } 4 + (6n)$$

Since the rules of operations order dictate that the multiplication comes before addition, the parentheses are optional.

b) The quotient of the number and five is decreased by three

$$\left(\frac{n}{5}\right) - 3 \text{ or } \frac{n}{5} - 3$$

Since division comes before subtraction, the parentheses are not needed.

There will be times when you don't need to place grouping symbols around a sum or a difference, but it is never wrong to place a sum or difference in parentheses, square brackets, or set brackets. For instance, the statement "four more than the sum of the number and six" is (n + 6) + 4 which equals n + 6 + 4 or n + 10.

Powers and roots frequently appear in stated problems. The problems in Example 4 will demonstrate some of the ways powers and roots may be involved.

## Example 4
Translate each of the given statements into an algebraic expression. Let n = the number.

a) The square of a number is increased by four

$$n^2 + 4$$

b) The square of the sum of the number and four

$$(n + 4)^2$$

c) The square of the number is added to the square of four

$$n^2 + 4^2$$

d) The square root of a number is increased by six

$$6 + \sqrt{n}$$

102

e)   The square root of the sum of the number and six

$$\sqrt{(n+6)} = \sqrt{n}+6$$

f)   The square of the number is decreased by the square root of the number

$$n^2 - \sqrt{n}$$

## Evaluating Algebraic Expressions

An algebraic expression may be evaluated in the same way that a formula is evaluated. We substitute the given value of each variable into that variable and then calculate. For example, evaluate:

$$3x^2 + 2x(4x - 5) + 3x - 1 \text{ when } x = -2$$
$$= 3(-2)^2 + 2(-2)(4(-2) - 5) + 3(-2) - 1$$
$$= 3(4) + 2(-2)(-8 - 5) + 3(-2) - 1$$
$$= 3(4) + 2(-2)(-13) + 3(-2) - 1$$
$$= 12 + 52 + (-6) - 1$$
$$= 12 + 52 + (-6) + (-1)$$
$$= 64 + (-7)$$
$$= 57$$

The preferred method for evaluating algebraic expressions is to simplify first and then evaluate the expression. Then:

$$3x^2 + 2x(4x - 5) + 3x - 1$$
$$= 3x^2 + 8x^2 - 10x + 3x - 1$$
$$= 11x^2 - 7x - 1$$

Substituting: $11(-2)^2 - 7(-2) - 1$
$$= 11(4) - 7(-2) - 1$$
$$= 11(4) - (-14) + (-1)$$
$$= 44 + 14 + (-1)$$
$$= 58 + (-1)$$
$$= 57$$

Please compare each method. Since there is less chance of making sign errors with the second method, that will be the method we shall use.

**Example 5**
Simplify as completely as possible and then evaluate each expression.

a)  $5x^2 - 4(x-2) + 3x(x-4)$  for x = -1

$= 5x^2 - 4x + 8 + 3x^2 - 12x$

$= 8x^2 - 16x + 8$

Then:

$8(-1)^2 - 16(-1) + 8$

$= 8 + 16 + 8$

$= 32$

b)  $3x^2 + 4x(x - y) + 2xy$ for x = -3 and y = -2

$= 3x^2 + 4x^2 - 4xy + 2xy$

$= 7x^2 - 2xy$

Then: $7(-3)^2 - 2(-3)(-2)$

$= 7(9) - 12$

$= 63 - 12$

$= 51$

# Exercises  2.5

In exercises 1 through 40 you are to translate the given statement into an algebraic expression. Let n represent the number. The answers may be found in the rear of the book.

1.   The number is increased by five

2.   Four more than the number

3.   Three less than the number

4.   Seventeen less the number

5.   Thirteen less twice the number

6.   Seven more than three times the number

7.   Eight less than one-half the number

8. Nine less than four times the number

9. Sixty-three less than five times the number

10. Three times the sum of the number and twelve

11. Four times the difference of the number and fifteen

12. One-fifth the sum of the number and twenty-one

13. Three times the difference of eleven and the number

14. One-sixth the product of five and the number

15. Six less than the square of the number

16. Six less double the number

17. Six less than the quotient of the number and four

18. Seven more than the 3rd power of the number

19. Three times the square root of the number

20. The square of the difference of the number and twelve

21. The square of the sum of the number and eight

22. Four times the square of the sum of the number and five

23. Three times the square of the difference of the number and sixteen

24. The square root of the sum of the number and six

25. The square root of the difference of the number and eleven

26. Six times the square of the number is decreased by five

27. One-fourth the square of the number is increased by seven

28. Twice the sum of one-half of the number and one-fourth

29. Four times the sum of one-fifth of the number and one-third

30. Six times the difference of three-fifths of the number and one-sixth

31. Five-elevenths of the number is decreased by three-fifths

32. The quotient of four and the number is increased by one

33. One less the quotient of one and the number

34. Six more than the product of seven and the number

35. Seven more than the product of the square of the number and two

36. Eleven less than the product of the square of the number and four

37. The sum of the square of the number and three times the number

38. The difference of sixteen and the square root of the number

39. The product of the sum of the number and six and the difference of the number and four

40. The product of the sum of the number and 5 and the sum of the number and 3

In exercises 41 through 60 you are to **simplify then evaluate** the given algebraic expression. On the test you **MUST SIMPLIFY FIRST** or you will not receive credit even if the final answer is correct. Show all work in the spaces provided. The answers may be found in the rear of the book.

41. $3x(2x - 1) + 4x^2$      for $x = -2$

42. $7x(2x - 3y) + 42xy - 3x^2$      for $x = -1$ and $y = -2$

43.  $2x(3x^2 - 4x - 7) + 3x^3 - 2x^2 + 4x - 12$  for x = -3

44.  $-2x(-4x + 3) - 5x(-2 - 7x) - 3x^2 - 4x - 5$     for x = -4

45.  $7(x - 3) + 4(x - 5) - 8(3x - 1) - 4x - 5$   for x = -1

46.  $6ax - \dfrac{a^2 x^2}{ax} + \dfrac{12x^4}{3x^3} - 5x$ for a = -2 and x = -5

47.  $2a(3a - x) + 4x(3x - a)$ for a = -1 and x = -2

48.  $5x^2(6 - x^2) + 2x^4 - 25x^2 + 3$ for x = 2

49.  $3x^2(4 - x) + 2x^3 - 5x^2 + 12$ for x = -5

50. $a^2(ab - c) - a^2c + 6a^3b$ for a $= 2$, b $= -3$, and c $= -5$

51. $3(x - y - z) - (x + y + z) - z$ for x $= -1$, y $= -2$, and z $= -3$

52. $2(a + b + c) - 4c + 3b - (2a - 5 - 3b)$     for a $= 2$, b$= 3$, and c $= 5$

53. $\dfrac{35x^3(x - y)^2}{7x^2(x - y)} - 3(x^2 + 4y)$ for x $= -4$ and y $= -1$

54. $2a^2 - 3b^2 - 4b - 3(3b - 1) - a^2$ for a $= 10$ and b $= -6$

55. $\dfrac{49x^5(3x - 5)^7}{-7x^4(3x - 5)^6} - \dfrac{81x^{11}(2x - 3)^8}{-9x^{10}(2x - 3)^7}$ for x $= -4$

56. $\dfrac{b}{b} - 2\dfrac{c^2}{c^2} + 7(4x - 5) - 12$ for $x = 3$

57. $5x - 3x^2 - 2x(4x - 7) - 5x + 13$ for $x = -4$

58. $3xyz - 4x^2yz - 2x(yz - 5xyz)$ for $x = 2$, $y = 5$, and $z = -3$

59. $\dfrac{4b^3(2b - 9)^3}{-2b^2(2b - 9)^2} + \dfrac{28b^4(7x - 5)^3}{-7b(7x - 5)^3}$ for $b = -2$

60. $5t^3 - 4t^2 - 3t + t(2t^2 - 4t - 5) - 17t - 5$ for $t = -1$

## 2.6   Review Exercises for Chapter 2

You will find this set of review exercises to be the largest set you have encountered. It was designed to give you an opportunity to practice what you have learned before you proceed to Chapter 3. Please take advantage of this opportunity.

### Vocabulary Exercises

1.  Define each of the following terms:

| | | |
|---|---|---|
| a)  formula | h)  algebraic term | o)  exponent |
| b)  general formula | i)  algebraic expression | p)  power |
| c)  specific formula | j)  like terms | q)  superscript |
| d)  literal symbols | k)  combining like terms | r)  subscript |
| e)  variable | l)  distributive law | s)  square |
| f)  coefficient | m)  distribution | t)  root |
| g)  constant | n)  base | u)  evaluation |

## Compare and Contrast Exercise

2. Compare and contrast arithmetic and algebra. Your discussion should include literal symbols, formulas, variables or unknowns, and other relevant terms.

In exercises 3 through 10 you are to identify the parts of each formula (coefficient, constant, or variable).

3. $A = 3R + 5S + 12$   4. $P = 3t - 4r + 7$   5. $T = -5q - 2p + 11$   6. $B = 6z + 2$

7. $Q = q + w - 8$       8. $D = 24 - T_2 + 6$   9. $h = 2s + 4$       10. $G = t - 3r + 4s$

In exercises 11 through 17 you are to translate the given statement into a general formula using the letters in parentheses.

11. The perimeter (P) of a triangle is the sum of the three side lengths (a, b, and c).

12. The circumference (C) of a circle is the product of pi ($\pi$) and the diameter (d).

13. The distance (D) traveled by a car moving at constant speed is the product of car's speed (s) and the time (t) of the drive.

14. The radius (r) of a circle can be found by dividing the circle's circumference (c) by the product of two and pi ($\pi$).

15. The temperature in degrees Celsius (C) is given by five-ninths of the difference of F and 32, where (F) is the temperature in degrees Fahrenheit.

16. The mass (m) of a body may be found by dividing the body's weight (W) by the acceleration due to gravity (g).

17. The volume (V) of a sphere is the product of four-thirds, pi ($\pi$), and the cube (3rd power) of the radius (r).

In exercises 18 through 23 you are to translate the given statement into a specific formula. Please use letters that will help your reader to follow your work. That is, use letters which suggest the variables they represent. For example, d suggests distance.

18. An office equipment company charges $90.00 per month plus 2¢ per copy to rent a top of the line photocopier. Write a formula for the cost of one year's rental.

19. A manufacturing firm that your company uses charges a $125.00 set-up fee and $17.50 per piece to make an aluminum chassis for the control panel for a bank of welding robots. Write a formula for the cost of any size production run for this part.

20. A computer software company packs 24 boxes of software in each carton and 12 cartons per pallet. Write a formula for the number of boxes (N) in a shipment of n pallets. Using the formula, find the number of boxes of software on 35 pallets.

21. The pay rate of a certain technician is $11.00 per hour with time and one-half for overtime (beyond 40 hours). Write a formula for the employee's weekly wage in terms of time. Using the formula find the wages for the technician if she worked 44 hours.

22. An out-of-town computer repair firm charges $100.00 for the first hour of a service call and $40.00 per hour for each additional hour to service your company's computers. Write a formula for their charge in terms of time. Using the formula find the labor charge for a four hour repair visit.

23. A corrugated paper company that supplies a company with shipping cartons charges a $75.00 set-up fee and $1.25 per piece to make a carton for a certain industrial electronic product the firm manufactures. Write a formula for the cost of any size production run for this carton. Using the formula, find the cost for a production run of 12,000 cartons.

In exercises 24 through 40 you are to evaluate the given formula at the specified values.

24. Evaluate the formula R = 3s + 3r + 12    given s = 4 and r = 3.

25. Evaluate the formula D = 120d - 48h + 36        given d = 3 and h = 5.

26. Evaluate the formula T = 16t² - 4t + 10    given t = -5.

27. Using the formula of problem 11 find the perimeter of a triangle with sides 3.45 in, 2.74 in and 4.15 in.

28. Using the formula of problem 12 find the circumference of a circle with a diameter that measures 7.5 cm.

29. Using the formula of problem 13 find the distance traveled by a car moving at 45 mi/hr for 2.5 hours.

30. Using the formula of problem 14 find the radius of the circle which has a circumference that measures approximately 25.13 in.

31. Using the formula of problem 15 find the temperature in degrees Celsius if the Fahrenheit temperature is 212 degrees.

32. Using the formula of problem 16 find the mass of a body (in slugs) if the body's weight is 400 lb and g = 32 ft/sec².

33. Using the formula of problem 17 find the volume of a sphere if its radius measures 4 in.

34. Using the formula of problem 18 find the cost of one year's rental if 90,450 photocopies were made during that year.

35. Using the formula of problem 19 find the cost of a production run for 5,500 pieces.

36. Electric resistance, measured in ohms ($\Omega$), is given by:

$$R = P/I^2.$$

Find the resistance if the power (P) is 700 (watts) and the current (I) is 11.5 A (amperes).

Note: Why do a problem like number 36 if you are not an electronics major? Answer: To prove to yourself that formula use is the same for every discipline.

37. The area of a circle is given by:

$$A = \pi r^2$$

Find the area of the circle with radius r = 16.5 cm.

38. The volume of a cone is given by:

$$V = \frac{1}{3} \pi r^2 h$$

Here r is the radius of the base and h is the height. Find the volume of the cone which has base radius 10 cm and height 18 cm.

39. The distance traveled by a car that begins at rest and accelerates constantly is given by:

$$d = \frac{at^2}{2}$$

Here a = the acceleration of the car and t = time. Find d if a = 12 ft/sec² and t = 4 sec.

40. The surface area A of a circular cylinder (a can) is given by:

$$A = 2\pi r(r + h)$$

Here r is the radius of the circular base and h is the height. Find the surface area of a cylindrical fuel tank with radius 6.0 ft and height 32.0 ft.

In exercises 41 through 78 you are to combine like terms where possible.

41. $5x + 2 + 3x + 6$

42. $2y - 7 + 7y + 5$

43. $4t - 5q + 7t + 11q$

44. $-3p - 5r + 8p + 9r$

45. $7x + 5 - 7x - 5$

46. $4x^2 - x + 12x - 2x^2$

47. $11x - 3 - 7x + 8$

48. $\frac{1}{4}x + 5 + \frac{3}{4}x - 4$

49. $8x^2 + 6x - 5x^2 - 12x$

50. $2xy - x + 6xy - 6x$

51. $12x + 8y - 5x + 3y - 7x + 9y$

52. $x + 2y - 3z + t - 4x + y - z + 6t$

53. $8p^2 + 4p - 3 + 5p^2 + 3p + 10$

54. $12y^2 + 24xy - 3x^2 - 9xy - 9y^2$

55. $11t^2 + 3t + 12t^2 + 7t + 5$

56. $13z^2 + 2z - 3xz + z - 3x + 4xz + 8x + 12z^2$

57. $10x^2 + 11x + 8 + 15x^2 - 8x + 15$

58. $14x^3 - 4x^2 + 8x^3 + 15x^2$

59. $16x^5 - 12x^3 + 14x^5 + 7x^3 - 57x$

60. $18t^2 - 12t + 7 + 8t^2 - 7t + 3 - t^2 + 12$

61. $4xy^2 + 15xy + 17xy^2 + 12xy + 8y^2 - 17y$

62. $ab + ab^2 - 2b^2 + a^2 + 7ab - 8b^2$

63. $cAT + 11T + 4cAT + 11A + 9cAT + 21T$

64. $xy^3 + 2xy^2 + xy + 12xy^2 + 8xy^3 + 11xy$

65. $T^2 - 2t^2 + 8T - 7t + 14T^2 + 9t^2 + T - 13t$

66. $7x^2 - 13x + 4 + 4x^2 + 3x - 15 + 5x^2 - 7x + 9 + 54x - 12$

67. $6r^2 + 11r - 5 + 12r^2 - 11r + 14r^2 - 13r + 4 - 8r$

68. $0.6x^2 - 0.15x + 3.3 + 1.2x^2 + 0.75x - 0.9$

69. $0.7y^2 + 0.96y - 0.25 + 1.7y - 0.6y^2 + 3.4$

70. $\dfrac{1}{5}x^2 - \dfrac{3}{7}x + \dfrac{4}{5}x^2 - \dfrac{3}{7}x$

71. $7x^2 + 0.25x - 1.45x^2 + 1.25x - 1.35$

72. $3t^2 - 0.7t + 1.5t^2 - 0.2t$

73. $12x^3y^2 - 8x^2y^2 + 7xy^2 - 6x^3y^2 - 5xy^2 + 14x^2y^2$

74. $8xy - 5xy^2 + 7x^2y + 8xy + 8x^2y + 15xy^2$

75. $\dfrac{3}{7}r^2 + \dfrac{2}{5}r + \dfrac{11}{7}r^2 - \dfrac{2}{5}r + 7$

76. $48x - 15x^2 + 64x + 25x^2 + 56x + 48x^2 - 16x - 14$

77. $2^2x^2 - 4^2x + 4^2x^2 + 5^2x - 21x^2 + 36x + 27$

78. $14x + 27y - 21z - 8x - 72y - 19z + 12x - 5y - 3z$

In exercises 79 through 113  you are to simplify as completely as possible.

79. $5x - 3 + 2(x + 2)$

80. $2(x + 3) + 7x - 5$

81. $3(7x - 5) + 5(8 - x)$

82. $-7(5x - 2) + 3x - 9$

118

83.   -4(7 - 3x) + 8x + 17

84.   -3(15x - 5) - 3(5x - 12)

85.   $8(5x^2 + 7x + 9) - 7x - 12$

86.   $5x^2 - 8(3x^2 + 7x + 9) - 2x + 11$

87.   $15x^2 - 9x - 4(5x^2 + 7x - 11) + 13x - 48$

88.   12a + 8b - 7c + 5(2a - 6b + 7c)

89.   6(x - 7) + 2(7 - x) + 8x + 3

90.   5x - 3y + 2(5x - y) + 5(y + 4y)

91.   -(5t - 12)

92.   9 - (5x - 7)

93.   2x + 9 - (2x + 9)

94.  $t + 7 - (t + 7)$

95.  $5x - 11 - (4x - 9)$

96.  $-(2x + 1) - (-2x - 1)$

97.  $3x - (x - 3) + 2x - 1$

98.  $5z - 1 - (12z - 11) + 5z - 8$

99.  $-(4x - 1) - (5x - 7) - (9 - 7x)$

100. $4x^2 - 7x - 9 - (3x^2 - 24x - 17)$

101. $8x^2 - 2x + 7 - (-5x^2 + 9x - 12)$

102. $6x^2 - 3xy + 9y^2 - 4(5x^2 - 7xy - 9y^2)$

103. $7a^2 - 15b + 12c - 5(3b - 2c - 7a^2)$

104. $1.8x - 9.3y + 2(1.8x + 0.1y)$

105.  $8.9x - 1.75 - 3(0.9x - 2.4)$

106.  $3.5(0.3x - 1.5y) - 4(0.2x - 1.2y)$

107.  $12(1.6x + 4.5y) - 2(3.3x - 1.5y)$

108.  $2.5t - 2.8s - 5(3.2t - 3.1s)$

109.  $8x^2 - 2.2x - 4.6 - 5(0.3x^2 + 1.1x - 2.4)$

110.  $-2(3t^2 - 3) - 5(-5t^2 + 7) + 12 + 4t^2 - 3t$

111.  $3x^2 - 2x^2 - 5z^3 - 4(5x^2 - 7y^2 - 2z^3 - 11)$

112.  $4(a - 2b) - 5(3a - 3b) - (-2a - 4b) + 2a - 5b$

113.  $0.3t - 0.4s - 3(0.5t - 0.4s) - 2(0.3t - 0.5s) + 1.2t - 0.7s$

In exercises 114 through 122 you are to perform the stated operations.

114.  Subtract three times the quantity $2x + 5$ from four times the quantity $7x - 9$.

115.  Subtract five times the quantity $3x^2 - 5x - 7$ from three times the quantity $7x^2 + 5x - 9$.

116. Subtract three times the quantity $4x^2 - 7x - 2$ from seven times the quantity $3x^2 + 5x - 2$.

117. Multiply the quantity $2x^2 + 3x - 5$ by $(-2)$ and add that product to the quantity $6x^2 - 5x + 11$.

118. Multiply the quantity $2x^2 - 8xy + 5y^2$ by $(-3)$ and add that product to the quantity $14x^2 - 3xy + 11y^2$.

119. Subtract the quantity $10x^2 + 5x - 3$ from the sum of the quantities $7x^2 + 4x + 3$ and $5x^2 + 7x + 9$.

120. Subtract the quantity $2x^2 - 9x + 3$ from the sum of the quantities $4x^2 + 9x + 7$ and $2x^2 + 5x + 13$.

121. Subtract three times the quantity $2x + 5$ from seven times the sum of the quantities $3x + 5$ and $4x - 3$.

122. Subtract five times the quantity $9x + 11$ from twice the sum of the quantities $2x - 7$ and $5x - 3$.

In exercises 123 through 157 you are to simplify as completely as possible.

123. $x(4x - 7)$

124. $-3x(3x + 2)$

125. $2x(7x - 2)$

126. $5x(7x - 9)$

127. $4x(2x^2 - 5x + 7)$

128. $-4x(5x^2 + 7x - 2)$

129. $3x(5x^2 + 7x - 4)$

130. $-2x(5x^2 - 8x - 2)$

131. $-3y(y^2 + 7y + 8) + 3y^2 + 5y - 9$

132. $2t^2(t - 11) + 5t(t^2 + 4t) + 6$

133. $4z(1 - 2z) + 8z(1 + 3z) - 14z^2 + 7z$

134. $5t(t^2 + 4t - 3) + 3t(5 - 2t + 5t^2)$

135. $(8x^4)(-4x^7)$

136. $(-2x^6)(7x^3)$

137. $(-2t^3)(-t^4)(-2t^5)$

138. $-5s^2(-4s^5)(-2)$

139. $(5a^3b^2)(-2a^2b^3)$

140. $(-2st^2)(-5s^2t)(-s^3)$

141. $8x(3x+y)+5y(x-2)+5x^2+4xy$

142. $-x^2(3x^2+2y-7z)+4z(x^2+7y)+5x^2z-3x^2y$

143. $-2x^2(3x-4xy-5y)+7x^3-8x^2y+9x^3y$

144. $\dfrac{25x^7}{-5x^3}$

145. $\dfrac{64r^5}{-4r^4}$

146. $\dfrac{256x^9}{32x^7}$

147. $\dfrac{48t^5}{16t^4} + 7t$

148. $\dfrac{15t^3r}{3t^2} + \dfrac{4r^2}{-2r} + 3r - 4$

149. $\dfrac{28a^3b^2}{7a^3b} + 2(3b - 5) + 2a - 4b$

150. $\dfrac{49x^2yz}{-7x^2z} - 4z(2 - z) + 7z^2 - 4z$

151. $\dfrac{2y^2}{2y^2} - \dfrac{5b^2y^2}{by} + b(4y) + 7 + 2by$

152. $\dfrac{12x(3x - 2)^3}{4x(3x - 2)^2} + \dfrac{27x^2(3x - 5)}{9x^2}$

153. $\dfrac{4c^2(3x+7)^7}{c^2(3x+7)^6} - 4(2x) + 3 + 12x - 1$

154. $\dfrac{128(2x-5)^3}{-64(2x-5)^2} + \dfrac{16s^2(2x+7)}{8s^2} + 4x + \dfrac{x^2}{x^2}$

155. $3R_1^2 + 4R_1 - 3R_1^2 + 15R_1$

156. $3T_1(-2T_1 + 3T_2) + 12T_1^2 + 3T_1T_2$

157. $3a_1^2(4a_1^3 + 7a_1^2 - 12a_1) - 2a_1^5 - 7a_1^3$

In exercises 158 through 174 you are to **simplify then evaluate** the given algebraic expression.

158. $4y(3y - 7) + 5y^2$ for $y = -1$

159. $5t(2t - 3r) + 11rt - 5t^2$ for $t = -1$ and $r = -4$

160. $4a(7a^2 - 5a - 7) + 8a^3 + 4a^2 + 4a - 15$ for $a = -3$

161. $-3x(7x + 8) + 4x(7x - 7) - 5x^2 + 4x - 9$ for $x = 5$

162. $-2(5x - 7) + 7(3x + 3) + 5(-2x - 9)$ for $x = -3$

163. $12x - \dfrac{3^2x^2}{3x} + \dfrac{15x^5}{3x^4} - 7x$ for $x = -5$

164. $4c(3c - 4x) + 5x(-2x - c)$ for $c = -3$ and $x = -2$

165. $12x^2 + 3x(-4x - 2) + 5x + 7$ for $x = 8$

166. $8x^2(5x - 4) + 7x^3 - 3x^2 + 4$ for $x = 3$

167. $t(tr - 3s) - t^2s + 4t^2r$ for $t = 2$, $r = 4$, and $s = -2$

168. $\dfrac{42x^4(3x - 7)^2}{-7x^3(3x - 7)} - 3(2x^2 + 5x) + 5x + 7$ for $x = -2$

169. $4x^2 - 7y^2 - 3y - 4(2y - 5) - 3x^2$ for $x = 4$ and $y = -3$

170. $\dfrac{36x^8(2x + 5)^3}{-4x^7(2x + 5)^2} + \dfrac{98b^4x(4x - 5)^5}{49b^4(4x - 5)^4}$ for $x = -2$

128

171. $\dfrac{2a}{a} - 5\dfrac{x^2}{x} + 3(2x + 3) - 5$ for $x = 4$

172. $5x^3 - 7x^2 - 2x(4x^2 - 7x - 5) - 8x + 9$ for $x = -2$

173. $\dfrac{48x^3(7x - 2)^3}{-24x^2(7x - 2)^2} + \dfrac{64x(3x - 4)^3}{-16x(3x - 4)^2}$ for $x = 3$

174. $8x^2yz - 4x^2y^2z - 2x^2(yz - 5y^2z)$ for $x = 1,\ y = 3,\ z = -2$

In exercises 175 through 208 you are to translate the given statement into an algebraic expression. Let n represent the number.

175. The number is decreased by four

176. Five less than the number

177. Three less than the number

178. Twelve less the number

179. Fourteen more than twice the number

129

180.  Six less than three times the number

181.  Seven less than twice the number

182.  Ten less than seven times the number

183.  Sixty-three less than five times the number

184.  Five times the difference of the number and twenty-three

185.  Six times the sum of the number and fifteen

186.  One-fourth the sum of the number and thirteen

187.  Five times the difference of eight and the number

188.  Three times the number is decreased by eight

189.  One-fifth the product of two and the number

190.  Seven more than the square of the number

191.  Eight more than the quotient of the number and four

192.  The sum of three and the number increased by seven

193.  The difference of six and the number decreased by five

194.  The square of the sum of the number and eleven

195.  Nine times the square of the difference of the number and fourteen

196.  Five times the square of the difference of the number and nine

197.  The square root of the sum of the number and three

198. The square root of the difference of the number and nineteen

199. Four times the square of the number is increased by twelve

200. One-seventh the square of the number is decreased by seven

201. Six times the sum of one-third of the number and one-seventh

202. One-half the sum of the number and one-fourth

203. Five times the difference of twice the number and four

204. Eleven times the sum of the number and six

205. Ten times the difference of four and the number is increased by seven

206. Sixteen less the product of five and the number

207. Eight more than the product of three and the number

208. Seven more than the product of the square of the number and two

# CHAPTER 3
## FIRST DEGREE EQUATIONS

INTRODUCTION:

In this chapter we begin our studies of equations with the simplest of all equations ...first degree equations. Although simple, first degree equations are powerful tools that will enable us to solve a vast array of problems.

OBJECTIVES:

Upon completion of this chapter you will be able to:

1. Explain what is meant by each of the terms printed in *italicized* type in this chapter.
2. Solve first degree equations.
3. Write and algebraically manipulate first degree literal equations and formulas.
4. Translate and solve word problems that involve first degree equations.
5. Write ratios and proportions and use proportions to solve relevant problems.

## 3.1 First Degree Equations

An *equation* is a statement that two quantities are equal. Familiar examples include:

$$2^3 = 8 \qquad \frac{32}{64} = \frac{1}{2} \qquad -3^4 = -81 \qquad 3(4+5) = 27$$

An *algebraic equation* is a statement that two algebraic expressions are equal. Examples include:

$$2x + 14 = 4x - 6 \qquad 7x^2 + 12x - 5 = 0 \qquad x^3 = 27 \qquad 4(2x + 5) = x^2$$

The *degree* of an equation is the highest sum of the powers of the variables in any one term of the equation. $x + 2 = 4$, $3x + y = 5$, and $4x - 5y = 7$ are first degree equations. $x^2 + 5x + 6 = 0$, $xy - 3 = 4y$, and $x^2 + y^2 = 4$ are second degree equations. $x^3 - 27 = 0$, $x^2y - 5 = 6$, and $xyz - xy = 8$ are third degree equations.

A *solution* of an equation is any value which when substituted into the equation yields a true statement. Sometimes it is said that the solution satisfies the equation.

## Example 1

Determine whether the specified value is or is not a solution of the given equation.

    a)  $2x - 4 = 10$; $x = 7$
        Substituting and evaluating:
        $2(7) - 4 = 10$

14 - 4 = 10

10 = 10 which is true, so 7 is a solution of 2x - 4 = 10.

b) 3x - 5 = 4; x = 5
   Substituting and evaluating:
   3(5) - 5 = 4
   15 - 5 = 4
   $10 \neq 4$, so 5 is not a solution of 3x - 5 = 4.

c) $x^2$ + 7x + 12 = 0; x = -3
   Substituting and evaluating:
   $(-3)^2$ + 7(-3) + 12 = 0
   9 + (-21) + 12 = 0
   0 = 0 which is true, so x = -3 is a solution of $x^2$ + 7x + 12 = 0.

d) $x^2$ + 3x = 7x; x = 2
   Substituting and evaluating:
   $(2)^2$ +3(2) = 7(2)
   4 + 6 = 14
   $10 \neq 14$, so 2 is not a solution of $x^2$ + 3x = 7x.

An equation is like a laboratory balance. If we begin with the same mass in each pan, we will have balance. We will not disturb the balance if we:

   1. add the same mass to both pans;
   2. subtract the same mass from both pans;
   3. mutiply the amount of mass in each pan by the same amount; or
   4. divide the amount of mass in each pan by the same amount.

Let's see how this applies to equations.

## The Four Properties of Equality

Property 1:   We may add any quantity to both sides of an equation without changing the solution of the equation.

   Example:  Solve x - 1 = 5
   Adding 1 to both sides of the equation gives
   x - 1 + 1 = 5 + 1 or
   x = 6

Property 2:   We may subtract any quantity from both sides of an equation without changing the solution of the equation.
   Example:  Solve x + 6 = 10

133

Subtracting 6 from both sides of the equation:
x + 6 - 6 = 10 - 6 or
x = 4

Property 3: We may multiply both sides of an equation by the same quantity (except zero) without changing the solution of the equation.

Example: Solve $\dfrac{x}{5} = 6$

Multiplying both sides of the equation by 5:

$5 \bullet \dfrac{x}{5} = 5 \bullet 6$ or

x = 30

Property 4: We may divide both sides of an equation by the same quantity (except zero) without changing the solution of the equation.

Example: Solve 3x = 12

Dividing both sides of the equation by 3 yields:

$\dfrac{3}{3}x = \dfrac{12}{3}$ or

x = 4

Each of the above equations required the use of only one of the four properties of equality to solve it. Most equations will be more involved and will require us to use two or more of the properties to solve them. Such equations are easily handled if we work in a step by step fashion.

**Example 2**
Solve each equation for the variable involved.

a) 2x - 7 = 17

Adding 7 to both sides of the equation (we always take care of the added or subtracted part first so that we don't create fractions):

2x - 7 + 7 = 17 + 7

or 2x = 24
Dividing both sides of the equation by 2:

$$\frac{2}{2}x = \frac{24}{2} \text{ or}$$

$$x = 12$$

b)  $3s + 6 = 21$

Subtracting 6 from both sides of the equation:

$3s + 6 - 6 = 21 - 6$ or

$3s = 15$

Dividing both sides of the equation by 3:

$$\frac{3}{3}s = \frac{15}{3} \text{ or}$$

$$s = 5$$

c)  $\frac{5}{3}t - 11 = 14$

Adding 11 to both sides of the equation:

$$\frac{5}{3}t - 11 + 11 = 14 + 11 \text{ or}$$

$$\frac{5}{3}t = 25$$

Multiplying both sides of the equation by 3:

$5t = 75$

Dividing both sides of the equation by 5:

$$\frac{5}{5}t = \frac{75}{5} \text{ or}$$

$$t = 15$$

Note : $\frac{5}{3}t = \frac{5t}{3}$ and may be written either way.

In our work we will encounter equations with variables on both sides of the equal sign. We use the same four properties of equality to solve such equations.

## Example 3

Solve each equation for the variable involved.

a) $5p - 2 = 3p + 18$

Subtracting 3p from both sides of the equation:

$5p - 2 - 3p = 3p + 18 - 3p$ or

$2p - 2 = 18$

Adding 2 to both sides of the equation:

$2p - 2 + 2 = 18 + 2$ or

$2p = 20$

Dividing both sides of the equation by 2:

$$\frac{2p}{2} = \frac{20}{2} \text{ or}$$

$p = 10$

b) $3x + 7 = 6x - 5$

Subtracting 3x from both sides of the equation:

$3x + 7 - 3x = 6x - 5 - 3x$ or

$7 = 3x - 5$

Adding 5 to both sides of the equation:

$7 + 5 = 3x - 5 + 5$ or

$12 = 3x$

Dividing both sides of the equation by 3:

$$\frac{12}{3} = \frac{3x}{3} \text{ or}$$

$$4 = x$$

In Chapter 2 we learned about combining like terms and the distributive rule. We shall now use combining like terms and the distributive rule as tools to help us solve equations.

## Example 4
Solve each equation for the variable involved.

    a)  $4(2r - 5) = 4r + 8$

        Distributing:

        $8r - 20 = 4r + 8$

        Subtracting 4r from both sides of the equation:

        $8r - 20 - 4r = 4r + 8 - 4r \text{ or}$

        $4r - 20 = 8$

        Adding 20 to both sides of the equation:

        $4r - 20 + 20 = 8 + 20 \text{ or}$

        $4r = 28$

        Dividing both sides of the equation by 4:

        $$\frac{4r}{4} = \frac{28}{4} \text{ or}$$

        $r = 7$

Note: It is customary to show only the result of the step rather than the operation involved. Following this custom, Example 4a would be written:

        $4(2r - 5) = 4r + 8$

        $8r - 20 = 4r + 8$

$$4r - 20 = 8$$

$$4r = 28$$

$$r = 7$$

b)  $3(2x - 6) + 2(10 + x) = 4x + 2$

Distributing on the left side of the equation:

$$6x - 18 + 20 + 2x = 4x + 2$$

Combining like terms on the left side of the equation:

$$8x + 2 = 4x + 2$$

Subtracting 4x from both sides of the equation:

$$4x + 2 = 2$$

Subtracting 2 from both sides of the equation:

$$4x = 0$$

Dividing both sides of the equation by 4:

$$x = 0$$

There will be times when we encounter equations containing fractions. These equations are easily changed to equations without fractions if we multiply both sides of the original equation by the least common denominator (LCD) of the fractional terms of the original equation.

## Example 5
Solve each equation for the variable involved.

a)  $\dfrac{x}{2} + \dfrac{x}{6} = 24$

Multiplying both sides of the equation by the LCD:

$$6\left(\frac{x}{2} + \frac{x}{6}\right) = 6 \bullet 24$$

Distributing on the left side:

$$6 \bullet \frac{x}{2} + 6 \bullet \frac{x}{6} = 6 \bullet 24 \text{ or}$$

$3x + x = 144$ or

$4x = 144$

Dividing both sides of the equation by 4:

$x = 36$

b) $\frac{t}{4} - 3 = \frac{t}{5} + 2$

Multiplying both sides of the equation by the LCD:

$$20\left(\frac{t}{4} - 3\right) = 20\left(\frac{t}{5} + 2\right) = 20 \bullet \frac{t}{4} - 20 \bullet 3 = 20 \bullet \frac{t}{5} + 20 \bullet 2 \text{ or}$$

$5t - 60 = 4t + 40$

Subtracting 4t from both sides of the equation:

$t - 60 = 40$

Adding 60 to both sides of the equation:

$t = 100$

c) $\frac{47}{60} = \frac{w}{12} + \frac{1}{5}$

Multiplying both sides of the equation by the LCD:

$$60 \bullet \frac{47}{60} = 60\left(\frac{w}{12} + \frac{1}{5}\right) \text{ or}$$

$47 = 5w + 12$

Subtracting 12 from both sides of the equation:

$$35 = 5w$$

Dividing both sides of the equation by 5:

$$7 = w$$

d) $\dfrac{r}{3} = \dfrac{r}{4} + 1$

Multiplying both sides of the equation by 12:

$$4r = 3r + 12$$

Subtracting 3r from both sides of the equation:

$$r = 12 \, .$$

We may occasionally be required to work with equations containing decimals. Let's study some examples of that style of equation.

## Example 6

Solve each equation for the variable involved.

a) $0.3(2x - 0.4) = 3x + 4.68$

Distributing on the left side of the equation:

$$0.6x - 0.12 = 3x + 4.68$$

Subtracting 0.6x from both sides of the equation:

$$-0.12 = 2.4x + 4.68$$

Subtracting 4.68 from both sides of the equation:

$$-4.80 = 2.4x$$

Dividing both sides of the equation by 2.4:

$$-2 = x$$

b) $3.2x - 4.1 = 2.7 - 4.7x$

Adding $4.7x$ to both sides of the equation:

$7.9x - 4.1 = 2.7$

Adding $4.1$ to both sides of the equation:

$7.9x = 6.8$

Dividing both sides of the equation by $7.9$:

$x \approx 0.86$

c) $0.9x + 0.5(10 - x) = 0.6(10)$

Distributing on the left side of the equation:

$0.9x + 5 - 0.5x = 0.6(10)$

Combining like terms on the left side and multiplying on the right side:

$0.4x + 5 = 6$

Subtracting 5 from both sides of the equation:

$0.4x = 1$

Dividing both sides of the equation by $0.4$:

$x = 2.5$

d) $0.6(5) + 0.2x = 0.35(5 + x)$

Multiplying on the left side and distributing on the right side:

$3 + 0.2x = 1.75 + 0.35x$

Subtracting $0.2x$ from both sides of the equation:

$3 = 1.75 + 0.15x$

Subtracting 1.75 from both sides of the equation:

$1.25 = 0.15x$

Dividing both sides of the equation by 0.15:

$8 \approx x$

## Types of Equations

There are three types of equations. These are conditional equations, identical equations, and contradictions. A *conditional equation* is an equation which is only true for certain values of its variable. For example, $2x - 3 = 7$ is a conditional equation because it is only true if $x = 5$. An *identical equation* or *identity* is true for any value of its variable. $3(2x - 5) = 6x - 15$ is an example of an identity since it is true no matter what value we substitute in for x. A *contradiction* is never true for any value of its variable. The equation $2(5x + 3) = 10x + 4$ is a contradiction because there is no x value that will make it a true statement.

If we try to solve identities or contradictions we will obtain results similar to those in Example 7.

### Example 7
Solve each equation.

      a)    $3(2x - 5) = 6x - 15$

          Distributing on the left side of the equation yields an equation with <u>identical</u> sides.

          $6x - 15 = 6x - 15$

          Subtracting 6x from both sides of the equation yields another identity.

          $-15 = -15$

          Adding 15 to both sides of the equation yields yet another identity.

          $0 = 0$

      b)  $2(5x + 3) = 10x + 4$

          Distributing on the left side of the equation, we obtain:

$$10x + 6 = 10x + 4$$

If we subtract 10x from both sides, we obtain the obviously false statement, that is, we obtain the contradiction:

$$6 = 4$$

# Exercises 3.1

Solve each equation for the variable involved. Some solutions may be fractions or decimals. Show all your work in the spaces provided. The answers may be found in the rear of the book.

1. $3x = 9$

2. $t - 1 = 6$

3. $s + 4 = 10$

4. $y/2 = 5$

5. $r + 5 = 7$

6. $2 - h = 6$

7. $w/5 = 4$

8. $16x = 48$

9. $3y - 1 = 8$

10. $4y + 6 = 14$

11. $5t + 6 = 21$

12. $2 - 3p = -7$

13. $-4 + 7q = -18$

14. $\dfrac{m}{4} - 1 = 11$

15. $\dfrac{z}{3} - 2 = 10$

16. $\dfrac{t}{-2} + 3 = -1$

17. $\dfrac{w}{-5} - 7 = 0$

18. $-3y - 5 = 1$

19. $-7 + 3p = -1$

20. $m - 5m = -16$

21. $3t - 7t = -20$

22. $8q - 5q = -27$

23. $3y = 7 - 4y$

24. $5 - 2x = 1$

25. $\dfrac{3x}{5} - 1 = 17$

26. $\dfrac{4x}{7} + 3 = 15$

27. $\dfrac{-5t}{3} + 7 = 13$

28. $-\dfrac{4}{5}p + 6 = 22$  Note: $-\dfrac{4}{5}p = \dfrac{-4p}{5} = \dfrac{4p}{-5}$

29. $-\dfrac{3}{7}r - 5 = -11$  Note: $-\dfrac{3}{7}r = \dfrac{-3r}{7} = \dfrac{3r}{-7}$

30. $9x - 4 = 7x$

31. $12x - 10 = 7x + 5$

32. $2t + 9 = 8t - 9$

33.  $12p + 11 = 5p + 4$

34.  $y - 4 + 2y = 12 + 9y - 36$     Reminder: Combine like terms on each side first.

35.  $2n + 5 + 2n = 3n - 8 + 7n - 1$

36.  $3t + 2(t + 1) = 12$

37.  $3(q - 5) = 3q + 2 + q$

38. $5(3d + 2) = 10d + 25$

39. $8r - 4 = 8 - 2(5 - r)$

40. $5 - 2(3p + 1) = 3(2 - 3p)$

41. $3(s + 1) - (s - 1) - (s - 1) = 4s + 2$

42. $8m + 2 = 2(9 - m) - 3 - (9 - m)$

43. $5s + 4(s + 20) = 500 - (s + 20)$

44. $15s + 15(s - 10) = 1,350$

45. $2w + 2(3w + 4) = 88$

46. $2w + 2(5w - 6) = 132$

47. $0.12x + 0.6 = 5.4$

48. $0.40x - 0.7 = 7.3$

49. $0.12x + 0.10(15,000 - x) = 1,600$

50. $0.09(11{,}000 - x) + 0.11x = 1{,}150$

51. $0.4y + 0.3(10 - y) = 3.5$

52. $0.7x + 0.4(20 - x) = 0.5(20)$

53. $0.35a + 0.2(8 - a) = 0.27(8)$

54. $0.5(8) + 0.9x = 0.8(8 + x)$

55. $0.6x + 0.2(2) = 0.45(2 + x)$

56. $\dfrac{2}{3}x + 4 = 2x - 4$

57. $\dfrac{4}{7}x + 2 = \dfrac{x}{2} + 3$

58. $\dfrac{5x}{3} - 5 = 2x - 9$

59. $\dfrac{x}{2} + \dfrac{4}{5} = \dfrac{4x}{5} - \dfrac{26}{5}$

60. $\dfrac{x}{3} + \dfrac{x}{5} = \dfrac{16}{15}$

61. $\dfrac{1}{2}x + \dfrac{1}{5}x + \dfrac{1}{7}x = x - 11$

In exercises 62 to 72 you are to determine whether the given equation is a conditional equation, an identical equation, or a contradiction.

62. $3(2x - 1) = 27$

63. $4(5x + 3) = 52$

64. $2(4x - 5) = 8x - 9$

65. $3(5x - 6) = 16 + 15x$

66. $3(2x - 5) + 4x = 10x - 15$

67. $6x + 17 = 7 - 2(-3x - 5)$

68. $3x - 6 = 3(x + 4)$

69. $2(4m - 1) + 3 = 5m + 4 + 3m$

70. $5(2x - 3) + 4(3x + 5) = 22x + 5$

71. $5(2x - 3) + 4(3x + 5) = 49$

72. $3x^2 - 6x + 12 = -3(-4 + 2x - x^2)$

## Review Exercises

73. Evaluate each of the given formulas at the specified values.

   a) $T = 3t - 4r + 6$ at $t = -2$ and $r = -5$

   b) $V = r^3 - 4\pi r^2 h$ at $r = 4$ and $h = 5$

   c) $T_2 = 4h_2 - 7k_2$ at $h_2 = 300$ and $k_2 = 4.25$

153

74. Simplify each of the following as completely as possible.

a) $4(3x - y) - 2(4x - 7y) - 5y + 7x$

b) $3x(7x + 4) - (x^2 - 3x) + 4x^2 - 17x - 12$

c) $V(V - R) - 3VR + 3V^2$

## 3.2 Manipulating Literal Equations and Formulas

A *literal equation* is an equation in which some or all of the constants are represented by letters. A *formula* is a literal equation relating two or more physical quantities. In this chapter section we learn how to manipulate literal equations and formulas. By *manipulation* we mean performing the operations needed to solve the literal equation or formula for a literal symbol other than the one for which it is currently solved. This work involves the solving techniques we learned in Section 3.1. Let's begin.

### Example 1

Solve the literal equation $y = 3x - c$ for x. Note solving for x also means to isolate x.

Adding c to both sides of the equation:

$$y + c = 3x$$

Dividing both sides of the equation by 3:

$$\frac{y + c}{3} = x$$

### Example 2

Solve the literal equation $y = 4x + b$ for b.

Subtracting 4x from both sides of the equation:

$$y - 4x = b$$

### Example 3

Solve the literal equation $y = 5x + k$ for x.

Subtracting k from both sides of the equation:

$$y - k = 5x$$

Dividing both sides of the equation by 5:

$$\frac{y - k}{5} = x$$

### Example 4

Solve the literal equation $y = mx + b$ for x.

Subtracting b from both sides of the equation:

$$y - b = mx$$

Dividing both sides of the equation by m:

$$\frac{y - b}{m} = x$$

## Example 5

Solve the literal equation 2x - 4y = ax for y.

Adding 4y to both sides of the equation (so the coefficient of y is positive):

$$2x = 4y + ax$$

Subtracting ax from both sides of the equation:

$$2x - ax = 4y$$

Dividing both sides of the equation by 4:

$$\frac{2x - ax}{4} = y$$

## Example 6

Solve the literal equation 2x - 4y = ax for x

**Because x appears in more than one term**, this problem requires us to use factoring. Put simply, factoring is breaking a product into its factors "unmultiplying". We might also say that factoring is undistributing.
Adding 4y to both sides of the equation:

$$2x = 4y + ax$$

Subtracting ax from both sides of the equation:

$$2x - ax = 4y$$

Factoring ( undistributing ) on the left side:

$$x(2 - a) = 4y$$

Dividing both sides of the equation by (2 - a):

$$x = \frac{4y}{2 - a}$$

## Example 7

Solve the literal equation $2(3x - 4y) - 2(3b + 5x) = 7x - 5y + 3b$ for y.

Distributing and simplifying on the left side of the equation:

$$6x - 8y - 6b - 10x = 7x - 5y + 3b$$

$$-8y - 6b - 4x = 7x - 5y + 3b$$

Adding 8y to both sides of the equation:

$$-6b - 4x = 7x + 3y + 3b$$

Subtracting 7x from both sides of the equation:

$$-6b - 11x = 3y + 3b$$

Subtracting 3b from both sides of the equation:

$$-9b - 11x = 3y$$

Dividing both sides of the equation by 3:

$$\frac{-9b - 11x}{3} = y$$

## Example 8

Solve the formula $C = 2\pi r$ for r.

Dividing both sides of the equation by $2\pi$:

$$\frac{C}{2\pi} = r$$

## Example 9

Solve the formula $A = \frac{1}{2} bh$ for b.

Multiplying both sides of the equation by 2 to clear the equation of fractions:

$$2A = bh$$

Dividing both sides of the equation by h:

$$\frac{2A}{h} = b$$

So far we have used only the four properties of equality, combining like terms, and the distributive rule to solve equations. Taking roots (square roots, cube roots, etc.) is yet another equation solving tool. See Examples 10 and 11.

## Example 10
Solve the formula $A = 2\pi r^2$ for r.

Dividing both sides of the equation by $2\pi$:

$$\frac{A}{2\pi} = r^2$$

Taking the square root of both sides of the equation:

$$\sqrt{\frac{A}{2\pi}} = r$$

## Example 11

Solve the formula $S = s_0 + \frac{1}{2}gt^2$ for t.

Subtracting $s_0$ from both sides of the equation :

$$S - s_0 = \frac{1}{2}gt^2$$

Multiplying both sides of the equation by 2 to clear the fractions:

$$2S - 2s_0 = gt^2$$

Dividing both sides of the equation by g:

$$\frac{2S - 2s_0}{g} = t^2$$

Taking the square root of both sides of the equation:

$$\sqrt{\frac{2S - 2s_0}{g}} = t$$

## Example 12
Solve the equation $T = k(T_2 - T_1)$ for $T_2$.

Distributing on the left side of the equation:

$$T = kT_2 - kT_1$$

Adding $kT_1$ to both sides of the equation:

$$T + kT_1 = kT_2$$

Dividing both sides of the equation by k:

$$\frac{T + kT_1}{k} = T_2$$

## Example 13
Solve the equation $S = \dfrac{d_1(S_2 - S_1)}{2d_2}$ for $S_1$.

Multiplying both sides of the equation by $2d_2$ :

$$2d_2 S = d_1(S_2 - S_1)$$

Distributing $d_1$ on the right side of the equation :

$$2d_2 S = d_1 S_2 - d_1 S_1$$

Adding $d_1 S_1$ to both sides of the equation :

$$2d_2 S + d_1 S_1 = d_1 S_2$$

Subtracting $2d_2 S$ from both sides of the equation :

$$d_1 S_1 = d_1 S_2 - 2d_2 S$$

Dividing both sides of the equation by $d_1$ :

$$S_1 = \frac{d_1 S_2 - 2d_2 S}{d_1}$$

Sometimes distribution is not needed. Example 14 demonstrates such a situation.

### Example 14
Solve $b = a(2c - 3)$ for a.

Dividing both sides of the equation by $2c - 3$ we obtain :

$$\frac{b}{2c - 3} = a$$

We need to distribute only if the variable we are solving for is inside the parentheses.

# Exercises  3.2

Solve the given literal equation or formula for the specified letter. Show all work in the spaces provided. The answers may be found in the rear of the book.
Note: Many of the following formulas come from technologies unrelated to yours. That does not matter...we need practice at solving formulas.

1.   $y = 2x - a$ for $x$

2.   $y = kx + 3$ for $x$

3.   $y = bx + c$ for $x$

4.   $y = A_1x + A_2$ for $x$

5.   $y - 2cx = 3b$ for $x$

6.   $y + 4kxz - 3 = 0$ for $x$

7. $ax - 3y + k = 3k$ for $y$

8. $3A_1x - 4A_2y = 2A_1$ for $y$

9. $y - 2x - c = 3x - 3c$ for $x$

10. $2ax - y = 3x + 4y$ for $x$

11. $2ky + 2x = 3y - x$ for $y$

12. $4y - 3bx = 6x + c$ for $x$

13. $A_1y + 7x = A_2y - 3x$ for $y$

14. $2(a + 3x) = y - 4x$ for $y$

15. $3a(4 + y) = 2x + 4a$ for $y$

16. $-4a(3x - by) = 2ax + 3aby$ for $y$

17. $0.5mx - 0.3y = -0.4m(0.2x + 0.7c)$ for $x$

18. $1.7x - 2.2b(2.7x - 1.3y) = 2.5x + 3.4bx$ for $y$

19. $S = \dfrac{d}{t}$ for $d$ ( physics )

20. $S = \dfrac{d}{t}$ for $t$ ( physics )

21. $z = \dfrac{x - \bar{x}}{s}$ for $\bar{x}$ ( statistics )

22. $P = a + b + c$ for $b$ ( geometry )

23. $C = \pi d$ for d  ( geometry )

24. $S = \dfrac{1}{2}(a + b + c)$ for a  $\left( \text{ geometry } \right)$

25. $B = MDT$  for D  (business)

26. $P = 2L + 2W$ for L     ( geometry )

27. $P = 240 + 0.03C$ for C ( payroll accounting )

28. $P = C(1 + m)$ for m ( business )

29. $\dfrac{P_1}{V_1} = \dfrac{P_2}{V_2}$ for $V_2$   ( chemistry, physics )

30. $K = \dfrac{1}{2} m v^2$ for m   ( physics, engineering )

31. $u = \dfrac{npv}{100 + v}$  for n   ( banking, business )

32. $P = 350 + 0.05C_1 + 0.1C_2$  for $C_1$   (payroll accounting )

33. $F_1 d_1 = F_2 d_2$ for $d_1$  ( physics )

34. $P = \dfrac{F}{A}$  for A  ( physics )

35. $W = F \bullet d$  for d  ( physics )

36. $d = \dfrac{m}{V}$  for V  ( physics )

37. P = BR for R ( business and other technologies )

38. P = BR for B ( business and other technologies )

39. $A = \frac{1}{2}h(b_1 + b_2)$ for $b_1$ ( geometry )

40. P = mgh for h ( physics )

41. $A = \frac{1}{2}ap$ for p ( geometry )

42. $g = \frac{2S}{t^2}$ for S ( physics )

43. I = prt for r ( banking )

44. $C = \pi d$ for $\pi$ ( geometry )

45. $A = p + prt$ for $p$ ( banking, business )

46. $F = \dfrac{9}{5}C + 32°$ for $C$ ( general science)

47. $V_1 = \dfrac{n_1}{n_2} V_2$ for $V_2$ ( chemistry )

48. $W = 40p + 1.5p(t - 40)$ for $t$ ( payroll accounting )

49. $A = \dfrac{\pi d^2}{4}$ for $\pi$ ( geometry )

50. $F_1 x = F_2(c - x)$ for $F_2$ ( physics )

51. $d = \dfrac{c - s}{n}$  for s  ( business )

52. $V = \pi r^2 h$ for h ( geometry )

53. $V = \pi r^2 h$ for r  ( geometry )

54. $LA = \pi r h_s$  for r  ( geometry )

55. $A = p + prt$  for t  ( banking )

56. $V = \dfrac{1}{3} \pi r^2 h$  for r  ( geometry )

57. $A = s^2$ for s   ( geometry )

58. $A = \dfrac{\pi d^2}{4}$ for d   ( geometry )

59. $K = \dfrac{1}{2}mv^2$ for v   ( physics )

60. $B = 180° - (A + B)$ for B   ( geometry, trigonometry, surveying )

61. $T = \dfrac{I}{PR} \bullet 360$  for R   ( business )

62. $\dfrac{a}{2} = \dfrac{b}{c}$  for c  ( geometry, trigonometry, algebra, and more )

63. $\overline{x} = \dfrac{a + b + c}{3}$  for b  ( statistics and other uses )

168

64. $Q = \dfrac{kA(T_2 - T_1)}{L}$ for $T_1$   ( physics )

65. $v = v_0 + at$ for a ( physics )

66. $V = \pi r^2 h$ for $\pi$  ( geometry )

67. $V = \pi r^2 h$ for r  ( geometry )

68. $A = p + prt$  for  r  ( banking )

69. $d = r(c - s)$  for c (business)

70  $m = p(1 + rt)$ for t (business)

71.  p = m(1 - dt) for m (business)

72.  p = m(1 - dt) for d (business)

73.  I = PRT for R  ( business )

74.  PV = nRT for  T  ( physics, chemistry )

75.  C + M = S   for C  (business)

76.  PV = nRT for  V  ( chemistry )

77.  $s = 3.87\sqrt{Rf}$  for R  ( traffic crash investigation)

78. $F = \dfrac{s^2}{30D}$ for D ( traffic crash investigation )

79. $s = \sqrt{30Df}$ for D ( traffic crash investigation )

80. $R = \dfrac{C^2}{8M} + \dfrac{M}{2}$ for C ( traffic crash investigation )

## **Review Exercises**

In exercises 81 through 96 you are to translate the given statement into an algebraic expression.

81. The number is decreased by seven

82. Eight less than the number

83. Ten less the number

84. Five less than twice the number

85. Ten more than three times the number

86. Four times the sum of the number and seven

87. Five times the difference of the number and ten

88. Three times the difference of nine and the number

89. Nine more than the square of the number

90. Four less than the quotient of the number and two

91. The product of two and the number decreased by seven

92. The square of the sum of the number and seven

93. Two times the square of the difference of the number and two

94. Four times the square of the number is increased by twelve

95. Three-sevenths the sum of the number and fourteen

96. Three times the difference of three times the number and five

### 3.3  A Most Useful Formula for Business and Industry

In this section we will study discounts, raises, commissions, interest, sales tax, miscellaneous increases and decreases, and other business and industry concepts. All of these topics can be handled using the simple percent formula we saw in Section 2.1:

$$P = B \bullet R = BR$$

To use this formula successfully we need the equation solving skills we learned in Section 3.1 as well as the formula manipulating skills we learned in Section 3.2. We will also have to learn to distiguish among the components of P = BR when we read a problem. That is, we must learn to identify which component is the **Part**, which component is the **Base**, and which component is the **Rate**. The following general rules will help you to distinguish among the components.

**Base:**          "is the starting point," "the whole of something." The base is often preceded by "of."

**Rate:**          a number followed by the percent symbol.

**Part:**          is always part of the base, often preceeded by "is."

**Note:** You must have two of the three components in order to solve for the third component.

## Example 1
During a recent shakedown at a corrections institution (cell and inmate inspection), a corrections officer stated that 15% of its inmates had illegal substances. If the corrections institution houses 15,000 inmates, find the number of inmates who possessed illegal substances.

First identify the Part, Base, and Rate.

Here P = number of inmates possessing contraband, B = total number of inmates, and R = percent of inmates possessing illegal substances.

Now evaluate the formula P = BR.

P = (15,000)(0.15)

P = 2,250 inmates had illegal substances.

## Example 2

A certain supermarket is required to collect 6.5% sales tax. If the store collected $500 in sales tax one day, what were the market's total sales for that day?

Identify the Part, Base, and Rate.

Here P = the sales tax, B = the total sales, and R = sales tax percentage.

Now evaluate the percent formula. Since 6.5% is 0.065 as a decimal we have

500 = B(0.065)

Dividing both sides of the equation by 0.065:

$7,692.31 = B

The total sales were $7,692.31.

## Example 3

Kim has $330 per month deposited into a stock purchase plan. If her annual earnings are $26,500 a year, what percent of her earnings does Kim have deposited into her stock plan?

Identify Part, Base, and Rate.

Here P = the amount invested, B = the annual earnings, and R = the percent of earnings deposited.

Please take note that the amount invested is given in months and the salary is given as an annual income. When you use the percent formula the units must match. So we change the monthly investment into an annual investement by multiplying by 12:

$330 • 12 would give a yearly amount of $3,960 invested.
Now we can use the percent formula to solve for percent of earnings.

3,960 = 26,500R

Dividing each side by 26,500:

0.149 = R

Kim has 14.9% of her earnings deposited into her stock fund.

## Example 4

A certain gas station sold 400,000 gallons of gasoline in March 1997 and 357,000 gallons in March 1998. Find what percent the sales decreased.

Identify the Part, Base, and Rate.

Here P = the decrease in gallons, B = last March's sales in gallons, and R = the percentage decrease.

We must first find the decrease in gallons before we can evaluate the percent formula.

Subtracting the number of gallons yields a decrease of 43,000 gallons.

Now we can evaluate the percent formula.

43,000 = 400,000R

Dividing both sides by 400,000:

0.108 = R

The percent of decrease from March 1997 to March 1998 was 10.8%.


## Example 5

A shipment of wine valued at $875 was damaged in transit. After inspection it was determined that 33% of the wine was damaged. What is the value of the undamaged wine.

Identify the Base, Rate, and Part.
Notice that we know the percent of damaged wine but we want to find the value of the undamaged wine. If the percent of damaged wine is 33%, then the percent of undamaged must be 100% − 33% = 67%.

Here P = the value of the undamaged wine, B = total value of all the wine, and R = the percenage of undamaged wine.

Now evaluate the formula.

P = (875)(0.67)

P = $586.25  is the value of the undamaged wine.

## Example 6

A police sergeant released a statement to the press that the number of hours worked rose from 150 hours last week to 210 hours due to overtime caused by the recent severe weather. Find the percentage increase in hours worked.

Identify the Part, Rate, and Base.

Since we are finding percent increase, we first have to find the increase in hours worked which would be 210 hours – 150 hours or 60 hours.

Here P = the increase in hours, B = last week's hours, and R = the percentage increase.

Now solve for percent increase.

60 = 150R

Dividing each side of the equation by 150:

0.4 = R

The percent increase in hours was 40%.

## Example 7

A certain computer manufacturer claims that 95% of its computers are free of defects. If 2,689 computers are found to be defective, find the total computers manufactured.

Identify the Part, Base, and Rate.

In this problem we are told the **percent** of undefective computers and the **number** of defective computers. When using the P = BR formula, the part and the percent must be expressed in the same unit. Therefore, we need to change the percent of undefective computers to the percent of defective computers. A simple calculation yields 100% – 95% = 5%.

Here P = the number of defective computers, B = the total number of computers, and R = the percentage of defective computers.

Now solve for the total computers manufactured.

2,689 = B(0.05)

Dividing both sides by 0.05:

53,780 = B

There were 53,780 computers manufactured.

## Exercises 3.3

Solve each problem using the percent formula (P = BR). Round to the nearest cent for money problems and to the nearest tenth if not money. Show all work in the spaces provided. The answers may be found in the rear of the book.

1.   A travel agent receives a commission of $275 for booking a trip to Asia. The total cost of the trip was $3,600. Find the percent the travel agent received.

2.   A massage shop owner spends 4.7% of her sales on advertising. If sales for the month were $9,275, find the amount spent on advertising.

3.   An owner of a pizza shop spends $600 a month on rent. If this is 30% of his sales, find the sales for the month.

4.   A computer priced at $1,225 sold for $1,050. Find the percent of price reduction.

5.   The average number of hours worked in a store last week increased from 42.3 hours to 45.7 hours. Find the percent increase in work hours.

6.    A stock increased 6.5% in value. If this caused a $5 increase, find the original price of the stock.

7.    A toy store received a shipment of goods with 3% of the toys damaged. If the value of the shipment was $2,300, find the value of the undamaged goods.

8.    A police officer saved 15% of her salary for a down payment on a house. This amounts to $475 a month. Find her annual salary.

9.    In an independent test, five out of 114 cars tested had engine problems. What percent of cars had problems?

10.   A manager of a store can use 1.5% of his total sales for maintenance. If his total sales are $730,000 this month, find the amount that may be spent on maintenance this month.

11.   A company estimates that it will spend $55,000 on shipping costs for the year. Total sales are estimated at $350,000. Find the percent of total sales that is spent on delivery.

12.   During a recent sale, a store offered a 25% discount on all toys in stock. Find the discount on a toy originally priced at $35.99.

13.　A computer programmer was making $50,000 a year until he received a raise. If he is now earning $53,000 a year, find the percent increase.

14.　In 1996 according to the NHTSA 17,182 people were killed in car crashes involving alcohol. This was 41% of all fatalities in the US involving car crashes. Find the total fatalities from car crashes. (Round to units place.)

15.　A restaurant owner states that 35% of her sales are from take out orders. If $2,700 of her sales are from dining room patrons, find her total sales.

16.　An office manager spends 15% of her budget on supplies. If her budget is $8,000, find the amount spent on supplies.

17.　A hotel manager states that 65% of the units revenue come from October through March. If the total revenue for the year was $580,000, find the sales for April to September.

18.　A correction institution states that 54% of its corrections officers have associates degrees. If the institution employs 298 officers, find the number of employees without degrees. (Round to units place.)

19.　A computer programmer spends 20% of her time debugging programs. If she works 48 hours a week, find the amount of time that she spends debugging programs each week.

20. A manager was given the challenge to reduce weekly employee hours from 170 to 150 per week. Find the percent decrease in employee hours.

21. A certain travel agent states that 16% of the airline tickets sold are for travel outside the United States. If total sales are $150,000, find the amount spent on travel inside the United States.

22. A certain hotel claims that it keeps 15% of its rooms for smokers. If the hotel has 675 rooms, how many are for smokers?

23. A certain restaurant usually discards 5% of its shipment of lettuce. If it has a shipment of 130 pounds of lettuce on a certain day, how many pounds of lettuce did the restaurant discard that day?

24. A certain manager of a store reports sales of $3,500 in December 1997 and $3,226 in December 1996. What was the percent of increase?

25. A certain clothing manufacturer claims that 23% of the items it produces are wholesaled as seconds. If $35,000 per month in sales are from seconds, find the total sales per month.

26. A salesman receives a commission of 3% of his sales. If he sold $35,000 in goods a certain month, find his commission for that month.

27. An office manager practiced her typing and increased her speed by 15 words per minute to 110 words per minute. Find the percent of increase.

28. A bed and breakfast owner's utilities bill increased from $150 in October 1996 to $174 in October 1997. Find the percent of increase.

29. A certain police chief announced that she was going to increase the number of hours that officers patrolled a certain highway from 20 hours a week to 32 hours a week. What is the percent of increase?

30. An automobile manufacturer claims that 88% of its cars are free of defects. If 375 autos were defective during a certain period, find the total number of cars manufactured during that period.

## 3.4 Applications of First Degree Equations in One Variable

In this chapter section we will translate and solve what are commonly called word problems or story problems. To do this we will use the equation-solving techniques that we learned in Section 3.1 together with much of what we have learned in the first two chapters.

It is true that some of what follows is difficult, but it can be mastered by the student who works wisely; that is, the student who works on this section regularly and often and who asks his instructor about any problem he is unable to solve.

Our work in this section has been divided into four parts. These are: general word problems; word problems involving motion; word problems involving mixtures; and word problems involving certain business or banking concepts.

### General Word Problems

*General word problems* are word problems that don't require any knowledge of geometry, physics, banking, or other special topics. We shall begin our studies of general word problems with several examples similar to the translation problems we worked in Chapter 2.

### Example 1
Twice the <u>sum</u> of a certain number and 5 is 22. Find the number.

> The first thing we do is read the problem thoroughly.
> The second thing we do is define the variable we are going to use:
> > Let n = the number
> Third, we translate the given information into an algebraic equation in terms of the variable we have chosen:
> > $2(n + 5) = 22$
> The fourth thing we do is simplify and then solve the equation:
> > $2n + 10 = 22$
> > $2n = 12$
> > $n = 6$
> The fifth and last task is to verify correctness by substituting the solution into the equation and evaluating: $2(n + 5) = 2(6 + 5) = 2(11) = 22$.

Whenever we solve a word problem we will follow these same five steps:
1. Read the problem thoroughly.
2. Define the variable. ( Let x be...)
3. Translate the problem into an equation using the variable we have chosen.

4. Simplify and solve the equation.
5. Verify that the solution is correct.

## Example 2

Two numbers add to 84. If the second number is 6 less than 5 times the first number, find the numbers.

Let n = the first number. Then 5n - 6 is the second number. Since the two numbers add to 84 we have:

$$n + 5n - 6 = 84$$

Simplifying and solving:

$$6n - 6 = 84$$
$$6n = 90$$
$$n = 15$$

Then the second number is 5n - 6 = 5(15) - 6 = 75 - 6 = 69

Verifying: 69 + 15 = 84                    **Answers: 15, 69**

## Example 3

One number is four more than twice a second number. If the two numbers add to 34, find the numbers.

Let n = the second number. Then the other or first number is 2n + 4. Since we are told that the two numbers add to 34 we have:

$$2n + 4 + n = 34 \text{ or } 3n + 4 = 34$$

Solving:

$$3n = 30$$
$$n = 10$$

Then the first number is 2n + 4 = 2(10) + 4 = 20 + 4 = 24

Checking: 24 + 10 = 34                    **Answers: 10, 24**

## Example 4

Three numbers add to 99. If the third number is 2 less than 4 times the first number and the second number is 5 more than three times the first number, find all three numbers.

Let n = the first number. Then the third number is 4n - 2 and the second number is 3n + 5. Since all three numbers add to 99 we have:

$$4n - 2 + 3n + 5 + n = 99$$

Simplifying and solving:

$$8n + 3 = 99$$
$$8n = 96$$
$$n = 12$$

Then the third number = 4n - 2 = 4(12) - 2 = 48 - 2 = 46, and the second number

is 3n + 5 = 3(12) + 5 = 36 + 5 = 41.
Verifying: 46 + 41 + 12 = 87 + 12 = 99.                    **Answers: 12, 41, 46**

## Note:

Consecutive integers are integers that differ by one. For example, 33 and 34 are consecutive integers. Odd integers that differ by two are called odd consecutive integers. Examples of odd consecutive integers include the pair 71 and 73. We will use this latter definition in our next example problem.

## Example 5

The sum of three consecutive odd integers is 129. Find the integers.

Let $x$ = the smallest odd integer. The middle integer is 2 more than the first integer, so it is $x + 2$. The third integer is 4 larger than the first so it is $x + 4$. Since all three numbers add to 129 we have:
$$x + x + 2 + x + 4 = 129$$
Simplifying and solving:
$$3x + 6 = 129$$
$$3x = 123$$
$$x = 41$$
Then the second is $x + 2 = 41 + 2 = 43$ and the third is $x + 4 = 41 + 4 = 45$.
Verifying: 41 + 43 + 45 = 129                    **Answers: 41, 43, 45**

## Example 6

A woman is eleven years older than twice her daughter's age. If the sum of their ages is 47, find each of their ages.

Let $d$ = the daughter's age. Then the woman's age is $2d + 11$. We are told that the sum of their ages is 47 so we have:
$$d + 2d + 11 = 47$$
Simplifying and solving:
$$3d + 11 = 47$$
$$3d = 36$$
$$d = 12$$
Then the woman's age is $2d + 11 = 2(12) + 11 = 24 + 11 = 35$
So the daughter's age is 12 and the woman's age is 35.
Verifying: 12 + 35 = 47                    **Answers: 12, 35**

## Example 7

A boy is three times as old as his brother. Three years ago he was six times as old as his brother. How old is each now?

Let $x$ = the younger boy's age now. Then $3x$ is the older boy's age now. Three years ago the younger boy was $x - 3$ years old and the older boy was $3x - 3$ years old. At that time the older boy was six times as old as the younger boy, so we have:

$$3x - 3 = 6(x - 3)$$

Simplifying and solving:

$$3x - 3 = 6x - 18$$
$$-3 = 3x - 18$$
$$15 = 3x$$
$$5 = x$$

Then the older boy's age is $3x = 3(5) = 15$

Verifying: Three years ago the younger boy was $5 - 3 = 2$, the older boy was $15 - 3 = 12$ and 12 is six times 2.                    **Answers: 5, 15**

## Note:

If we went into a hardware store and purchased 12 bolts at 13¢ each and 15 bolts at 18¢ each, then our total purchase (without sales tax included) is $12 \cdot 13¢ + 15 \cdot 18¢ = 156¢ + 270¢ = 426¢$ or $4.26. In general, it's quantity times price plus quantity times price plus... This type of thinking will be used in the next two example problems.

## Example 8

A purchasing manager placed an order for two types of motors. Given that one type of motor costs $2.40 more than the other type and that the manager spent $126 for 5 of each type of motor ( 10 total ), find the cost of each type of motor.

Let $c$ = the cost of the less expensive motor. Then $c + 2.40$ is the cost of the more expensive motor. Since five of each were purchased, we have:

$$5c + 5(c + 2.40) = 126.00$$

Simplifying and solving:

$$5c + 5c + 12 = 126.00$$
$$10c + 12 = 126.00$$
$$10c = 114.00$$
$$c = 11.40$$

Then the cost of the more expensive motor is $c + 2.40 = 11.40 + 2.40 = $13.80$.
Verifying: $5 \cdot \$11.40 + 5 \cdot \$13.80 = \$57.00 + \$69.00 = \$126$.  **Answers: $11.40, $13.80**

## Example 9

A purchasing manager ordered 5,000 batteries from one supplier and 3,000 of the same battery from a second supplier. She would have ordered all the batteries from the first supplier because their price is 8¢ per battery lower than the second supplier's price, but they only had 5,000 in stock. If both orders totaled $5,680.00, find each supplier's price for a single battery.

Let $p$ = the higher price or second supplier's price for the battery. Then $p - 8 =$ the first supplier's price. We have:

$$5,000(p - 8) + 3,000p = 568,000 \quad \text{( in pennies )}$$

Simplifying and solving:
$$5,000p - 40,000 + 3,000p = 568,000$$
$$8,000p - 40,000 = 568,000$$
$$8,000p = 608,000$$
$$p = 76¢$$

Then the first supplier's price is p - 8 = 76¢ - 8¢ = 68¢.

Verifying: $5,000•68¢ + 3,000•76¢ = 340,000¢ + 228,000¢ = 568,000¢ = \$5,680.00.$

**Answers:  Second supplier charges 76¢ per battery**

**First supplier charges 68¢ per battery**

## Note:

Many problems involving rental fees for equipment, hourly pay, or labor charges for repairs or service calls are similar to the above purchase problems. Examples 10 and 11 demonstrate such problems.

## Example 10

An office equipment company charges $90.00 per month plus 2¢ per copy to rent a top-of-the-line photocopier. If the charges for one year's photocopier rental was $3,990.50, how many copies were printed that year?

Let n = the number of copies made that year. Then:
$$0.02n + 12(90.00) = 3,990.50$$

Simplifying and solving:
$$0.02n + 1,080.00 = 3,990.50$$
$$0.02n = 2,910.50$$
$$n = 145,525$$

Verifying: $\$0.02(145,525) + \$1,080.00 = \$2,910.50 + \$1,080.00 = \$3,990.50$

**Answer: 145,525 copies**

## Example 11

An electrician charges $25.00 per hour for her time and $15.00 per hour for her assistant's time. The assistant worked on a job for one hour and then was joined by the electrician. If the total labor bill for that job was $135.00, how many hours did the electrician and her assistant work <u>simultaneously</u> on that particular job?

Let t = the time the electrician worked on the job. The assistant worked one hour more, so her work time was t + 1. Then:
$$25.00t + 15.00(t + 1) = 135.00$$

Simplifying and solving:
$$25.00t + 15.00t + 15.00 = 135.00$$
$$40.00t + 15.00 = 135.00$$
$$40.00t = 120.00$$
$$t = 3 \text{ hours}$$

Then the assistant worked t + 1 = 3 + 1 = 4 hours

Verifying:  $\$25.00•3 + \$15.00•4 = \$75.00 + \$60.00 = \$135.00$    **Answer:  3 hours**

## Example 12

At a bank a man handed a teller a twenty dollar bill and asked for change in quarters and dimes. If he received ten more quarters than dimes, how many of each coin did he receive?

Let d = the number of dimes the man received. Then d + 10 is the number of quarters he received. Putting the number of coins together with their values yields:
$$10d + 25(d + 10) = 2{,}000 \quad \text{( in pennies )}$$
Simplifying and solving:
$$10d + 25d + 250 = 2{,}000$$
$$35d + 250 = 2{,}000$$
$$35d = 1{,}750$$
$$d = 50$$
Then there were d + 10 = 50 + 10 = 60 quarters.

Verifying: $50 \cdot 10¢ + 60 \cdot 25¢ = 500¢ + 1{,}500¢ = \$5.00 + \$15.00 = \$20.00$.

**Answers: 50 dimes, 60 quarters**

## Example 13

A young boy has been saving change for several months. If he has 21 more quarters than half dollars, 52 more dimes than quarters, 10 fewer nickels than dimes, no pennies, and his coins are worth $51.70 total, find how many of each coin he has.

You should notice that the number of each kind of coin is related to the number of half dollars. So let x = the number of half dollar coins. Then the number of quarters is x + 21, the number of dimes is x + 21 + 52 = x + 73, and the number of nickels is x + 73 - 10 = x + 63. Putting the number of coins together with their values yields:
$$50x + 25(x + 21) + 10(x + 73) + 5(x + 63) = 5{,}170 \quad \text{( in pennies )}$$
Simplifying and solving:
$$50x + 25x + 525 + 10x + 730 + 5x + 315 = 5{,}170$$
$$90x + 1{,}570 = 5{,}170$$
$$90x = 3{,}600$$
$$x = 40$$
Then there are x + 21 = 40 + 21 = 61 quarters, x + 73 = 40 + 73 = 113 dimes, and x + 63 = 40 + 63 = 103 nickels.

Verifying: $40 \cdot 50¢ + 61 \cdot 25¢ + 113 \cdot 10¢ + 103 \cdot 5¢ = 2{,}000¢ + 1{,}525¢ + 1{,}130¢ + 515¢ = \$20.00 + \$15.25 + \$11.30 + \$5.15 = \$51.70$.

**Answers: 40 half dollars, 61 quarters, 113 dimes, 103 nickels**

## Example 14

An oil drum 1/3 full was found to be 4/5 full after 14 gallons were added. Find the capacity of the drum.

Let C be the capacity of the drum. Then:

$$\frac{1}{3}C + 14 = \frac{4}{5}C$$

Then multplying through by the LCD 15 we have:

$$15 \bullet \frac{1}{3}C + 15 \bullet 14 = 15 \bullet \frac{4}{5}C$$

$$5C + 210 = 12C$$
$$210 = 7C$$
$$30 = C$$

Verifying:

$$\frac{4}{5}(30) - \frac{1}{3}(30) = 24 - 10 = 14$$

**Answer : 30 gallons**

## Example 15

If three more than one-fifth a number is equal to one-half the number, find the number.

Let n = the number. Then we have:

$$\frac{1}{5}n + 3 = \frac{1}{2}n$$

Distributing through by the LCD 10 we have:

$$10 \bullet \frac{1}{5}n + 10 \bullet 3 = 10 \bullet \frac{1}{2}n$$

$$2n + 30 = 5n$$
$$30 = 3n$$
$$10 = n$$

Verifying:

$$\frac{1}{5} \bullet 10 + 3 = \frac{1}{2} \bullet 10 \Rightarrow 2 + 3 = 5$$

**Answer: 10**

## Motion and Rate Word Problems

The speedometer in a car tells us the speed of the car in miles per hour or kilometers per hour. Generalizing, average speed is distance divided by time or $s = d/t$.

## Example 16
Using the formula $s = d/t$ calculate the desired quantity:
  a)  Find the speed of a plane if it flew 600 kilometers north in 3 hours.
  b)  Find the distance traveled by a train moving at an average speed of 35 mi/hr for 7.5 hr.
  c)  Find the time it took a man to drive 450 km if his average speed was 75 km/hr.

Part a)  $s = \dfrac{d}{t} = \dfrac{600\text{km}}{3\text{hr}} = 200\,\dfrac{\text{k m}}{\text{hr}}$

Part b)  Solving $s = \dfrac{d}{t}$ for d yields $d = st$.

Then $d = 35\,\dfrac{\text{mi}}{\text{hr}} \bullet 7.5\text{hr} = 262.5\text{mi}$

Part c)  Solving $d = st$ for t yields $t = \dfrac{d}{s}$.

Then $t = \dfrac{450\text{km}}{75\,\dfrac{\text{km}}{\text{hr}}} = 6\text{ hr}$

## Example 17
Two cars originally 700 mi apart are traveling toward each other at average speeds of 45 mi/hr and 55 mi/hr, respectively. If they maintain their average speeds, how long will it take them to cover the distance?

This an example of what some mathematicians call a convergence problem, since the two cars are converging on each other. The key to this (or a divergence problem) is that both cars are helping to cover the distance. That is, the distance covered by the slower car plus the distance covered by the faster car equals 700 miles. Let $t$ = the time of travel. Then using $d = st$ we have:

$$45t + 55t = 700$$

189

Simplifying and solving:
$$100t = 700$$
$$t = 7 \text{ hr}$$

**Answer: 7 hours**

## Example 18

A sales representative drove from her office to a customer's location in 4 hours. On the return trip the rep experienced heavy traffic and had to drive 15 mi/hr slower. If the return trip took 5.5 hours, find the round trip distance traveled by the sales rep.

The key to this motion problem is that the distance is the SAME or equal both ways.

Let s = the speed going to the customer's location. Then the speed on the return trip was s - 15. Using d = ts and setting the two distances equal, we have:
$$4s = 5.5(s - 15)$$
Simplifying and solving:
$$4s = 5.5s - 82.5$$
$$-1.5s = -82.5$$
$$s = 55 \text{ mi/hr}$$
Since the time going to the customers was 4 hr, the distance going was (55 mi/hr)•4 hr = 220 mi and the round trip distance is twice that or 440 mi.

Verifying: s - 15 = 55 - 15 = 40 and 55•4 = 40•5.5 = 220 mi.
**Answer: 440 miles**

## Example 19

A helicopter flying at 90 mi/hr departed at 10 a.m. A turboprop plane flying at 275 mi/hr began to pursue the helicopter at 11:30 a.m. How long will it take the plane to catch the helicopter?

This an example of a pursuit problem. The key to a pursuit problem is that both vehicles must cover the SAME distance. Let t = the time of flight for the turboprop plane. Then, since the helicopter took off one and one-half hours earlier, its time of flight is t + 1.5. Using d = ts and setting the distances equal we have:

$$275t = 90(t + 1.5)$$
Simplifying and solving:
$$275t = 90t + 135$$
$$185t = 135$$
$$t \approx 0.73 \text{ hr or about 44 min}$$
Verifying: (Due to rounding we are approximating)

$$275 \frac{mi}{hr} \cdot 0.73hr \approx 201mi \text{ and } 90\frac{mi}{hr}(0.73hr + 1.5hr) = 90\frac{mi}{hr} \cdot 2.23hr \approx 201mi.$$

**Answer: 0.73 hours or 44 minutes**

Many rate problems are similar to motion problems. Example 22 demonstrates a manufacturing production scheduling problem.

## Example 20
A manufacturer of stamped metal products has two machines scheduled to be used for a certain production run. If the new machine can stamp 720 parts per hour while the older machine has a rate of 420 parts per hour, how long will it take both machines to produce 34,560 parts if the new machine is set-up and started one-half hour before the older machine?

Let t = the time that both machines are working similtaneously. Then the newer machine will run for t + 0.5 hours. Then we have:
$$420t + 720(t + 0.5) = 34,560$$
Simplifying and solving:
$$420t + 720t + 360 = 34,560$$
$$1,140t + 360 = 34,560$$
$$1,140t = 34,200$$
$$t = 30 \text{ hr}$$

Verifying: $420\frac{parts}{hr} \cdot 30hr + 720\frac{parts}{hr}(30hr + 0.5hr) = 12,600 \text{ parts} + 21,960 \text{ parts}$

$= 34,560 \text{ parts}.$

**Answer: 30 hours**

## Word Problems Involving Mixtures

We begin with a most useful model...the two item or two component mixture model where the final amount is specified.

## Example 21
A clerk needs 50 lbs of a candy mix that is 40% peppermints. If she has a candy mix that is 30% peppermints and another that is 70% peppermints, how much of each must be mixed to make the 50 lbs of a mix that is 40% peppermints?

## Discussion Break
We need to discuss a tool we will use frequently. It is an algebraic way to say "the difference". If we purchase a dozen donuts of which 8 are glazed, then 12 - 8 = 4 (the difference) are not glazed. If we purchase 100 feet of rope and cut off 80 feet, then 100 - 80 = 20 feet (the difference) of rope remains.
Generalizing: If we purchase a dozen donuts of which x are glazed, then 12 - x

(the difference) are not glazed. If we purchase 100 feet of rope and cut off x feet, then 100 - x feet (the difference ) of rope remains. Now back to the example problem.

Let x = the weight of 70% peppermint mix used. Then 50 - x (the difference) = the weight of 30% peppermint mix used. ( They must sum to 50 lbs since that is the goal) Then, relating the peppermint concentration:
$$0.70x + 0.30(50 - x) = 0.40(50)$$
Simplifying and solving:
$$0.70x + 15 - 0.30x = 20$$
$$0.40x + 15 = 20$$
$$0.40x = 5$$
$$x = 12.5 \text{ lb of the 70% peppermint mix}$$
Then 50 - x = 50 - 12.5 = 37.5 lb of 30% peppermint mix must be used.
Verifying: 0.70(12.5) + 0.30(37.5) = 8.75 + 11.25 = 20 (which is 40% of 50)
**Answers: 12.5 lb, 37.5 lb**

A variation of the two component mixture problem is a mixture where the final total is not specified, but the amount of one of the components of the mixture is specified. The next example problem demonstrates such a mixture problem.

## Example 22
How much pure water must be added to 3 quarts of 30% acetic acid vinegar to dilute it down to 10% acid?

Here the volume (3 quarts) of one of the components has been specified and the final volume has not. The question is this: How many quarts of 100% water must be added to 3 quarts of 70% water (30% acid) to make 90% water?

Let x = the number of quarts of 100% water added. Then relating the percent of water:
$$0.70(3) + 1.00x = 0.90(3 + x) \qquad \text{Note: } 100\% = 1.00$$
Simplifying then solving:
$$2.10 + 1.00x = 2.70 + 0.90x$$
$$2.10 + 0.10x = 2.70$$
$$0.10x = 0.60$$
$$x = 6 \text{ quarts of 100% water}$$
Verifying: Out of the 3 qt, 70% or 2.1 qt was water. Now adding 6 quarts of pure water brings the total to 9 qt of which 2.1 qt + 6 qt = 8.1 qt is water. Then 8.1 qt/9 qt = 0.9 so, we have 90% water or 10% acid.
**Answer: 6 quarts**

## Word Problems Involving Concepts from Business or Banking

Some problems involving purchases may be solved using the same approach as the the two-component mixture problem with a specified total.

### Example 23

A purchasing manager ordered two different sizes of computer disks from a supplier. If the 3.5-inch disks sold for 70¢ each, 5.5-inch disks sold for 90¢ each, and the purchasing manager spent $4,600.00 for 6,000 disks, how many of each size disk did she purchase?

Let x = the number of 5.5-inch or 90¢ disks purchased. Then 6,000 - x = the number of 3.5-inch or 70¢ disks purchased. Writing a purchase equation similar to Example 8 we have:

$$0.90x + 0.70(6,000 - x) = 4,600.00$$

Simplifying then solving:

$$0.90x + 4,200 - 0.70x = 4,600.00$$
$$0.20x - 4,200 = 4,600.00$$
$$0.20x = 400$$
$$x = 2,000 \text{ of the 5.5-inch disks}$$

Then 6,000 - x = 6,000 - 2,000 = 4,000 of the 3.5-inch disks.

Verifying: $0.70(4,000) + $0.90(2,000) = $2,800.00 + $1,800.00 = $ 4,600.00.

**Answers: 4,000 3.5-in disks, 2,000 5.5-in disks**

## Simple Interest

Let r = annual interest rate, P = principal, and I = earnings. Then the earnings on an account held for one year are given by I = Pr. For example, the earnings on $1,000.00 invested at 8% simple interest for one year are: I = ($1,000.00)(0.08) = $80.00. We shall use simple interest in the next example problem.

### Example 24

A company invested $40,000.00 surplus funds in two accounts and had total earnings of $4,200.00. If one account lost 12% while the other earned 18%, find the amount invested in each account.

Let x = the amount invested at 18%. Then 40,000.00 - x was invested at -12% (The loss). Then:

$$0.18x + (-0.12)(40,000.00 - x) = 4,200.00$$

Simplifying then solving:

$$0.18x - 4,800.00 + 0.12x = 4,200.00$$

$$0.30x = 9{,}000.00$$
$$x = \$30{,}000.00$$
Then $\$40{,}000.00 - x = \$40{,}000.00 - \$30{,}000.00 = \$10{,}000.00$.
Verifying: $0.18(\$30{,}000.00) + (-0.12)(\$10{,}000) = \$5{,}400.00 - \$1{,}200.00 = \$4{,}200.00$.

**Answers: $30,000 at 18%, $10,000 at -12%**

Many times we may be called upon to use percentages to solve for past price or cost or to predict a future price or cost. Examples 27 and 28 demonstate such situations.

## Example 25

A new car with identical equipment costs your company 18% more than it did three years ago. If the truck sells for $17,970.94 now, what was the new cost of the company's three year-old car?

Let x = the car's cost three years ago.
Since the car now costs 18% more, it sells for 118% of x. Then:

$$1.18x = 17{,}970.94$$
So $\qquad\qquad x = \$15{,}229.61$

**Answer: $15,229.61**

## Example 26

If new truck prices have climbed an average of 3.5% each of the last five years, what should a truck that now sells for $22,560.00, cost three model years from now?

This is a bit tricky since we will have mark-up on mark-up. That is, we can't simply multiply 3.5% by 3 and take off from there.

At the end of the first year the cost would be 103.5% of $22,560.00 = $23,349.60.
At the end of the second year the cost would be 103.5% of $23,349.60 = $24,166.84.
At the end of the third year the cost would be 103.5% of $24,166.84 = $25,012.68.

Could this be done by formula? The answer is yes. A formula relating the initial price to a fixed annual percentage increase is:

$$C = c(1+p)^t$$

where C is the future cost, c = the current cost, p is the annual percentage increase (as a decimal), and t is the number of years.

Let's use the formula to handle the calculations.

$$C = 22{,}560.00(1+0.035)^3 = 22{,}560.00(1.035)^3 = 22{,}560.00(1.1087179) = \$25{,}012.68$$

# Exercises 3.4

You are to use the five steps outlined between Example problems 1 and 2 to solve each of the following problems. You must use algebra...trial and error or guessing are not satisfactory methods. Show all work in the spaces provided. The answers may be found in the rear of the book.

**STOP** Do not attempt to do problems 1 through 22 until you have read and studied Examples 1 through 15.

## General Word Problems

1.   Twice the <u>difference</u> of a certain number and 7 is 46. Find the number.

2.   Two numbers add to 88. If the first number is 8 more than 9 times the second number, find the numbers.

3.   One number is seven less than three times a second number. If the two numbers add to 73, find the numbers.

4.   Three numbers add to 91. If the third number is 2 less than 3 times the first number and the second number is 3 more than twice the first number, find all three numbers.

5.   The sum of three consecutive even integers is 342. Find the integers.

6. A man is five years younger than three times his brother's age. If the sum of their ages is 35, find each of their ages.

7. A 16 ft long counter comes in two sections. If the one section is 4 ft longer than the other section, find the lengths of the two sections.

8. A woman is now five times as old as her daughter. In six years the woman will be three times as old as her daughter is then. How old is each now?

9. A computer technician placed an order for two different types of printer cartridges. Given that the cheaper cartridge costs $5.60 less than the more expensive cartridge and that the technician spent $160.00 for 4 of each cartridge (8 total), find the cost of each cartridge.

10. The manager of your hotel ordered two different styles of lamps for each room of your hotel. If each desk lamp cost $28.15 less than each floor lamp, and the manager spent $15,275.00 for one hundred lamps of each style (200 total), find the cost of one of each style lamp.

11. A travel agency ordered 8,000 brochures from one supplier and 6,000 brochures from a second supplier. They would have ordered all the brochures from the second supplier because their price is 10¢ per brochure lower than the first supplier and they ship freight prepaid, but they only had 6,000 brochures in stock. If both orders totaled $8,100, of which $300.00 was the first supplier's shipping charge, find each supplier's price for a single brochure.

12. A telephone company charges $150.00 per month plus 20¢ per minute for an in-state toll free (800) phone service for your company. If the charge for one month's service was $1815.30, how many minutes of calls were received that month?

13. The shipping division of your company has three clerks who process orders. During an average hour the middle seniority clerk handles 3 packages less than the senior clerk and 5 more than the new clerk. If all three clerks combined handle an average of 49 packages per hour, find how many packages each clerk ships per hour.

14. Three divisions of your company reported their gross sales. Division I's sales were four times those of division III and three times those of division II. If the gross sales for all three divisions were $34,000,000.00, find the gross sales of each division.

15. An accountant charges $25.00 per hour for her time and $12.50 per hour for each of her three bookkeeper's time. If two of the bookkeepers worked on a business tax return for nine hours and were then joined by the accountant, and the total bill for that job was $375.00, how many hours did the accountant and her two bookkeepers work simultaneously on that tax return?

16. Your company called in a computer repair service to clean and check the company's pc's and its mainframe. The service technician's assistant came at 8 a.m. to open and prepare the machines for the service technician, who arrived at 9:30 a.m. and they both finished together. The charge for the assistant's time is $20.00 per hour and the charge for the service technician's time is $40.00/hr. If the labor charges were $180.00, how long did the technician work on that visit?

17. Ticket sales for a police equipment show totaled $15,318.00. Advance tickets cost $5.00 each while tickets purchased at the door cost $6.00 each. If the number of $6.00 tickets sold was 3 more than twice the number of advance sale tickets, how many $6.00 tickets were sold?

18. A man cashed his paycheck at his bank. The teller handed him some twenty-dollar bills, some ten-dollar bills, some five-dollar bills, some one-dollar bills, two dimes, and 4 pennies. If the teller gave him the same number of tens as twenties, two more fives than twenties, and half as many ones as twenties, and his paycheck was $365.24, how many twenties did the teller give him?

19. A collection of coins contains half dollars, quarters, and dimes. If there 40 more dimes than quarters and 18 less halves than quarters and the collection has a face value of $122.50, determine how many of each coin there are in the collection.

20. A refrigerated wine dispenser keg 1/4 full was found to be 3/5 full after 17.5 liters of white wine were added. Find the capacity of the keg.

21. A man bought a used car and wanted to know the capacity of the gas tank. He stopped to buy gas when the gauge read 1/4 full. If 15 gal of gas moved the gauge to 7/8 full, what is the capacity of the car's gas tank?

22. If seven less than three-fifths of a number is equal to one-fourth the number, find the number.

**STOP**  Do not attempt to do problems 23 through 32 until you have read and studied Examples 16 through 20.

## Word Problems Involving Motion or a Rate

23.  Using the formula $s = d/t$ calculate the desired quantity:

   a) Find the speed of a ship if it sailed 60 miles east in 5 hours.
   b) Find the distance traveled by a motorcycle moving at an average speed of 60 km/hr for 3.25 hr.
   c) Find the time it took a plane to fly 700 mi if its average air speed was 120 mi/hr.

24.  Two ships originally 1,000 nautical miles apart are traveling toward each other at average speeds of 12 knots ( 1 knot = 1 nautical mile/hr ) and 8 knots respectively. If they maintain their average speeds, how long will it take them to cover the distance?

25.  A horseback rider traveling at 8 mi/hr departed at 7 a.m. Another rider departed the same location at 8 a.m. traveling at 12 mi/hr on the same trail as the first rider. How long will it take the second rider to catch the first rider?

26.  A jet flying east at 550 mi/hr in calm air passes a jet flying west at 500 mi/hr at 1:32 p.m. Assuming neither jet changes direction, at what time will the jets be 1,225 mi apart?

27. On a round trip flight a small plane flew with the wind going out and against the wind on the return trip. If the trip going out took 4 hr, the return trip took 6 hr, and the wind was blowing 30 mi/hr, find the round trip distance.

28. A courier drove from a warehouse to a retail outlet in 4 hr. On the return trip he experienced heavy traffic and his average speed was 10 mi/hr less than his average speed going. If the return trip took 5 hours, what was his average speed driving to the retail outlet?

29. A vice president of Quality Inns drove to and from a hotel in seven hours. If she drove 45 mph to the hotel and 60 mph on the way back, find the round trip distance.

30. A plastic molding plant has two presses scheduled to be used for a certain production run. If the machine with a twelve mold turret can produce 1,800 pieces per hour while the machine with an eight mold turret has a rate of 1,200 pieces per hour, how long will it take both machines to produce 72,000 pieces if the 8 mold machine is available for production 6 hours before the 12 mold machine?

31. An experienced assembler can assemble 12 components per hour on average. An apprentice assembler can assemble 9 components per hour on average. If three experienced assemblers begin work at 8 a.m. and are joined at 10 a.m. by four apprentice assemblers, how long will it take the group to finish assembling a run of 360 components?

32. Three computer controlled lathes have been scheduled to fill a production run of spindles. The production rates for the three lathes are 20 spindles/hr, 15 spindles/hr, and 12 spindles/hr. If the fastest lathe starts at 7 a.m., the next fastest lathe starts at 8 a.m., and the slowest lathe starts at 8:30 a.m., at what time can the assembly department pick-up a load of 120 spindles?

**STOP** Do not attempt to do problems 33 through 39 until you have read and studied Examples 21 and 22.

**Word Problems Involving Mixtures**

33. A chef needs 20 lbs of a coffee that has 20% arabica beans. If she has a coffee that is 50% arabica beans and another that contains 10% arabica beans, how many pounds of each must be mixed to make the 20 lbs of 20% mix?

34. How many liters of pure alcohol must be added to 3 L of 20% alcohol in water to make 50% alcohol in water?

35. A certain 2-cycle string trimmer used by the hotel groundskeeper requires a 1 to 16 mix of oil and gas. If you have a 10% oil in gas mix, how much pure gasoline must be added to make 2 gal of the 1 to 16 mix? **Hint**: A 1 to 16 mix means 1 of oil to 16 of gas or 1 part of 17 is oil...what is 1/17 as a percent?

36. A candy mix that sells for $2.50/lb is to be mixed with a candy mix that sells for $5.00/lb. How much of each should be used to make 50 lb of mix that sells for $4.00/lb?

37. How much of a candy that is 10% butterscotch should a clerk mix with 10 kg of mix that is 32% butterscotch to make a mix that is 24% butterscotch?

38. A nut mixture is 20% cashews. How many kg of cashews must be mixed with 10 kg of the 20% cashew mix to produce a mix that is 35% cashews?

39. In a certain state a restaurant is required to use 3% bleach to sterilize dishes. How many gallons of water must be mixed with 20 gallons of 5% bleach to make the 3% bleach?

**STOP** Do not attempt to do problems 40 through 47 until you have read and studied Examples 23 through 26.

## Word Problems Involving Concepts from Business or Banking

40. An investment broker invested $10,000.00 for you in two accounts. One account earned 9% while the other earned 7%. If the combined earnings were $830.00, how much was invested in each account?

41. A company invested $3,500,000.00 in two new divisions. One of the new divisions did very well, earning 12%, while the other division experienced a loss of 7%. If the combined gross earnings for both divisions was $135,000.00, how much was invested in each division?

42. A new delivery truck with identical equipment costs your company 13% more than it did 2 years ago. If the truck sells for $27,240.57 now, what was the new cost of the company's two year-old truck?

43. Use the formula of Example 26 to predict the cost of a new car 5 years from now if the current model is $18,750.00 and the price has undergone an average increase of 2.8% per year for the last six years.

44. You invested $35,000 in two accounts. One account earned 6% while the other earned 15%. If the combined earnings were $4350, how much was invested in each account?

45. A company invested $7,000,000 in two new divisions. One of the new divisions did very well, earning 16%, while the other division experienced a loss of 9%. If the combined gross earnings for both divisions was $245,000, how much was invested in each division?

46. Use the formula of Example 26 to predict the cost of a new car 7 years from now if the current model is $22,500 and the price has undergone an average increase of 3.7% per year for the last 6 years.

47. A new tractor with identical equipment costs your company 16% more than it did five years ago. If the tractor sells for $21,000 now, what was the new cost of the company's five year old tractor?

## 3.5 Ratio & Proportion

A *ratio* of one quantity to another quantity is the first quantity divided by the second quantity. The ratio of a to b is then a/b. It is common practice to use a colon as an alternative to the fraction bar. Then the ratio of a to b is written a:b.

### Example 1
Write the following ratios in each of the three possible forms.

a) The ratio of 15 to 7 is 15/7 or 15:7.

b) The ratio of 10 mi to 4 mi is 10 mi/ 4 mi = 5 mi/ 2 mi = 5/2 = 5:2.

c) The ratio of 1 to 16 is 1/16 or 1:16.

d) A boy weighs 112 lb and is 64 in tall. His weight to height ratio is 112 lb/ 64 in  or 7 lb/ 4 in or 1.75 lb/ 1 in.

e) A plane flys 440 mi in 2 hr. Its speed or distance to time ratio is 440 mi/2 hr or 220 mi/ 1 hr.

f) A map specifies that 1 in = 40 mi. Its map distance to road distance ratio or scale is then 1 in/ 40 mi or 1 in:40 mi.

g)  A weight of 20 lb stretches a certain spring 2.5 in. The spring's weight to length ratio or stretch ratio is 20 lb / 2.5 in or 8 lb / 1 in or 8 lb:1 in.

h) A mixture of bourbon in water is 2 parts bourbon to 7 parts water or 2 parts/ 7 parts or 2/7 or 2:7.

Sometimes it is desirable to change one of the units so that both parts of the ratio have the exact same unit. After the reduction the units will cancel leaving us with a numerical ratio. See Example 2.

### Example 2
a) The ratio 1.5 ft to 4 in is $\dfrac{1.5 \text{ ft}}{4 \text{ in}} = \dfrac{18 \text{ in}}{4 \text{ in}} = \dfrac{9}{2}$

b) The ratio 8 fl oz to 2 qt is $\dfrac{8 \text{ fl oz}}{2 \text{ qt}} = \dfrac{8 \text{ fl oz}}{64 \text{ fl oz}} = \dfrac{1}{8}$

A statement of equality between two ratios is called a *proportion*. Proportions are very useful mathematical tools. Example 3 demonstrates proportions and how they are solved.

## Example 3

a) $\dfrac{x}{20} = \dfrac{7}{10}$

This proportion is read "x is to 20 as 7 is to 10."

To solve the proportion multiply both sides by 20.

$20 \cdot \dfrac{x}{20} = 20 \cdot \dfrac{7}{10}$ or $x = 14$

b) $\dfrac{18}{x} = \dfrac{9}{11}$

This proportion is read "18 is to x as 9 is to 11."

Multiplying both sides by x:

$18 = \dfrac{9x}{11}$

Multiplying both sides by 11:

$198 = 9x$

Dividing both sides by 9:

$22 = x$

c) $\dfrac{x}{4,500 - x} = \dfrac{2}{3}$

This proportion is read "x is to 4,500 - x as 2 is to 3."

Multiplying both sides by 4,500 - x:

$x = \dfrac{2(4,500 - x)}{3}$

Multiplying both sides by 3 and distributing on the right side:

$3x = 9,000 - 2x$

Adding 2x to both sides:

$5x = 9,000$

Dividing both sides by 5:

$x = 1,800$

Now let's examine some applications of proportions.

## Example 4

A powdered mix used in a certain recipe calls for 3 parts of mix to 8 parts of milk. How many parts of milk must you add to 9 parts of the mix?

Let x = the number of parts milk to be added.

$$\frac{3}{8} = \frac{9}{x}$$ Then $3 \bullet x = 8 \bullet 9$ or $3x = 72$, so $x = 24$ parts milk.

## Example 5

A certain sanitizing mix requires a soap water to bleach ratio of 16:1. If a can contains 2.2 gallons of soap water, how many ounces of bleach must be added to obtain the 16 to 1 mix?

We must first reduce gallons to ounces so that both volumes are expressed in the same unit.

$$2.2 \text{gal} \bullet \frac{128 \text{ fl oz}}{1 \text{ gal}} \approx 280 \text{ fl oz}$$

Let x = the volume in ounces of bleach added.

Then $\dfrac{x}{280} = \dfrac{1}{16}$

Multiplying both sides by 280:

$$x = \frac{280}{16} = 17.5 \approx 18 \text{ fl oz}$$

## Example 6

Five out of every seven households have cable TV. If 40,000 households in a certain city have cable TV, how many do not have cable TV?

Let x = the number of households.

$$\frac{5}{7} = \frac{40,000}{x}$$ Then $5x = 280,000$, so $x = 56,000 =$ the number of households.

Then 56,000 - 40,000 = 16,000 households don't have cable TV.

Sometimes we will be given a sum of both parts rather than one of the parts. Example 7 demonstrates such a problem.

**Example 7**

Two machines making the same part produced 1,160 parts during a production run. If the faster machine produced five parts for each three parts the slower machine produced, find the number of parts produced by each machine during the run.

Method 1: Let x be the number of parts produced by the faster machine. Then the slower machine produced the remainder or 1,160 - x.

Then $\dfrac{x}{1,160 - x} = \dfrac{5}{3}$

Multiplying both sides by 3:

$\dfrac{3x}{1,160 - x} = 5$

Multiplying both sides by 1,160 - x:

$3x = 5(1,160 - x)$

Simplifying and solving for x:

$3x = 5,800 - 5x$

$8x = 5,800$

$x = 725$ and $1,160 - x = 1,160 - 725 = 435$

Method 2:

A ratio of 5 to 3 means that the faster machine produced 5 out of every 8 parts and that the slower machine produced 3 out of every 8 parts.

Therefore: $\dfrac{5}{8} \bullet 1,160 = 725$ and $\dfrac{3}{8} \bullet 1,160 = 435$

**Example 8**

A woman is 25 years old and her son is 5 years old. How many years will it take until the ratio of the mother's age to the son's age is 2:1?

Let x = the number of years until the ratio is 2 to 1.

$\dfrac{25 + x}{5 + x} = \dfrac{2}{1}$  Then $1(25 + x) = 2(5 + x)$.  So $25 + x = 10 + 2x$ or $15 = x$.

# Exercises 3.5

Please show all work in the spaces provided. The answers may be found in the rear of the book. In exercises 1 to 14 you are to write the given ratio in fraction form. Remember to reduce where possible.

1. 13 to 5    2. 21 to 7    3. 6 to 14    4. 18:10    5. 35 to 21    6. 84 to 49    7. 9:33

8. 45 to 9    9. 63 to 54    10. 33:22    11. 36 to 27    12. 144:108    13. 18:6    14. 3:21

In exercises 15 to 26 you are to write the given ratio in fraction form. Remember to reduce where possible. Eliminate units if both units represent the same physical quantity (see Example 2).

15. 2 m to 40 cm        16. 1.5 lb to 4 oz        17. 32 lb to 12 in$^2$        18. 0.42 Ω to 50 ft

19. 4 in to 14 lb        20. 4 L to 60 sec        21. 3 mi to 1,320 ft    22. 750 lb to 2.25 ton

23. 40 sec to 3 min        24. 96 lb to 4 ft$^3$        25. 2.4 kg to 300 mg    26. 1.25 in to 10 lb

In exercises 27 to 34 you are to write and then reduce the stated ratios.

27. Average speed is the ratio of distance to time. Find the average speed of an object that traveled 792 ft in 9 sec.

28. The mass density of a substance is the ratio of the mass of the substance to the volume of the substance. If 36.32 cm$^3$ of an alloy has a mass of 748.35 g, find the density of the alloy.

29. The flow rate of a gas or liquid is the ratio of the volume of the gas or liquid to time. If 62,400 ft$^3$ of gas moves in 52 min, find the flow rate of the gas.

30. Aeronautics engineers use Mach numbers to describe the speed of fast aircraft. The Mach number is the ratio of the plane's speed to the speed of sound (740 mi/hr). Find the Mach number of a plane flying at 1,650 mi/hr.

31. Automobile fuel efficiency is the ratio of the volume of gasoline used to the distance driven. If a truck consumed 34 gallons of gas during a 758.2 mile trip, find its fuel efficiency.

32. Pressure is the ratio of force to area. Find the pressure exerted by a 360.8 lb object sitting on 14.3 in² area.

33. Specific gravity is the ratio of the density of a material to the density of water. If the density of aluminum is approximately 165.5 lb/ft³, the density of water is 62 lb/ft³, find the specific gravity of aluminum.

34. If a company has $4,019,200 in assets and $1,256,000 in liabilities, find the company's assets to liabilities ratio.

In exercises 35 to 42 you are to solve the given proportions.

35. $\dfrac{x}{4} = \dfrac{27}{12}$

36. $\dfrac{81}{x} = \dfrac{27}{42}$

37. $\dfrac{x}{55} = \dfrac{12}{11}$

38. $\dfrac{3}{5} = \dfrac{x}{25}$

39. $\dfrac{25}{35} = \dfrac{10}{x}$

40. $\dfrac{x}{12 - x} = \dfrac{1}{3}$

41. $\dfrac{x}{6,400 - x} = \dfrac{3}{5}$

42. $\dfrac{125 - x}{x} = \dfrac{3}{2}$

In exercises 43 to 69 you are to use proportions to solve each problem.

43. If a jet travels 1,620 ft in 2 sec, how far will it travel in 27 sec?

44. According to the Bureau of Justice Statistics there were (in 1993) 21 fulltime local police officers for every 10,000 citizens. Using this ratio as a guide, how many full-time officers should be employed by a city with a population of 27,500?

45. If a new plastic molding machine produces 7 parts for each 5 produced by the old machine, how many parts will the old machine produce while the new machine makes 2,870 parts?

46. If the property tax rate in a certain county is $18.55 per thousand dollars of property value assessment, find the property tax on a home assessed at $72,000.

47. If a mail sorting machine sorts 200 pieces of mail in 2.5 min, how long will it take the machine to sort 680 pieces?

48. If the ratio of rum to cocktail mix in a certain drink is 2 to 5, how many ounces of mix are needed if one ounce of rum is used?

49. A certain laser printer prints 15 pages in 2.5 min. How long would it take the printer to print 2 pages?

50. If a company's workforce is composed of 2 full-time to 5 part-time employees and the company has 420 full-time employees, how many part-time employees does the company have?

51. If a watch is gaining 1.5 min each 8 hr, how many minutes will it gain in 3.5 days?

52. If the ratio of inmates to correction officers at a certain correctional institution is 65 to 2, how many inmates does the institution house if there are 73 COs?

53. A certain gasoline powered generator runs 1.5 hr on 1 gal of gasoline. How long would the generator run on 3.5 gal of gasoline?

54. The ratio of females to males at a certain college is 53 to 45. If the enrollment of the college is 3,626, how many female students are enrolled?

55. Minnesota is sometimes referred to as the land of 10,000 lakes. If in a certain region of Minnesota the ratio of lake surface area to land surface area is 3 to 5 and the total acreage of the region is 32,000 acres, find the number of acres of lakes in the region.

56. If a watch is losing 1.5 min each 12 hr, how many minutes will it lose in one week?

57. If an operator of a new computer controlled lathe produces 5 pieces for each 3 pieces he produced on a standard lathe, find his expected shift output on the new machine if his expected shift output was 48 on the standard machine.

58. If the incidence of a disease in a certain city is 3 cases per 10,000 people, how many people in the city have the disease if the total population of the city is 605,784?

59. A computer manufacturer sells both desk-top and portable computers in the ratio of 11 to 3. If last year's sales of both types combined was 858,200 units, how many portable computers were sold?

60. If the ratio of single occupancy rooms to double occupancy rooms at a certain hotel is 2 to 9 and the hotel has 220 rooms, how many double occupancy rooms does the hotel have?

61. If the ratio of weight to volume ( density ) of a certain substance is 2,220 lb to 4 ft.$^3$, how many cubic feet would 1,300 lb of the substance occupy?

62. A certain inclined section of a highway has a rise to run ratio of 1.5 in to 20 ft. How many inches does the highway rise over a 450 ft run?

63. A company has two cash reserve accounts. If the ratio of the two accounts is 5 to 2 and both accounts together total $14,000,000, how much money is in each account?

64. If a certain ink jet printer has an ink to copy ratio of 20 mL to 300 copies, how many copies may be run with a 4 fl oz bottle of ink?

65. If three out of every five households in the U.S. have home computers. how many households in a city with a population of 135,000 probably don't have a home computer?

66. On a drawing the drafter specified the scale as 1.5 in to 8 ft. Find the length of a beam if it is 3.75 in long on the drawing.

67. On a certain map the scale is given as 2 in = 30 mi. A technician measured some distances on the map. If the total distance she measured on the map was 27.5 in, find the road distance.

68. A company invested $25,000 in one account and $15,000 in another account. The company then made an equal second investment in each account. If this second investment made the ratio of the two accounts 5:4, find the amount added to each account.

69. A man is 30 years old and his son is 10 years old. How many years will it take until the ratio of the father's age to the son's age is 2:1?

## 3.6 Review Exercises for Chapter 3

Show all work (where required) in the spaces provided. The answers may be found in the rear of the book.

### Vocabulary Exercises

1. Define each of the following terms:

    a) equation                         h) contradiction

    b) algebraic equation              i) literal equation

    c) degree of an equation          j) formula

    d) solution of an equation       k) ratio

    e) solving an equation            l) proportion

    f) conditional equation

    g) identical equation

In exercises 2 through 26 you are to solve the given equation for the variable it contains.

2. $\dfrac{4x}{5} + 11 = 19$

3. $\dfrac{-4s}{3} + 7 = 15$

4. $17 - \dfrac{4}{5}t = 1$

5. $2x - 4 = 6x$

6. $4x - 7 = 7x + 5$

7. $3s + 6 = 8s - 9$

8. $-5t + 11 = 5t - 9$

9. $2x - 5 + 6x = 4x + 7$

10. $3x + 5(2x + 3) = 28$

11. $5(y - 2) = 3y + 14$

12. $12 + 6(-3z + 8) = 3(2 + 3z)$

13. $-4(x + 1) - (x - 2) - 2(x - 1) = x - 16$

14. $7x + 2(x - 5) = -2(3x - 25)$

15. $0.70y + 0.60 = 3.40$

16. $1.2x - 0.7 = 4.1$

17. $0.70x + 0.45(10 - x) = 0.50(10)$

18. $0.07(20,000 - x) + 0.8x = 1,450$

19. $0.92x + 0.38(20 - x) = 0.56(20)$

20. $0.3(12) + 0.5x = 0.44(12 + x)$

21. $0.75x + 0.35(10) = 0.60(10 + x)$

22. $\frac{3}{5}x + 4 = x$

23. $\frac{1}{2}x + 4 = x$

24. $6(3x - 1) = -7(8 + x)$

25. $3(x + 1) = 4(6 - x)$

26. $2(x + 1) - 3(4x - 2) = 6x$

In exercises 27 to 34 you are to determine whether the given equation is a conditional equation, an identical equation, or a contradiction.

27. $4(3x - 1) = 32$

28. $-4(5x + 2) = 72$

29. $7(2x - 5) = 35 + 14x$

30. $-2(5x - 6) = 12 - 10x$

31. $-4(2x - 5) - 2x = 20 - 10x$

32. $3(2x^2 - 4x + 5) = 15 + 2x(3x + 6)$

33. $-3x(-2x - 6) + 7 = 8 + 6x(x + 3)$

34. $-3x(-2x - 6) + 7 = 8 + 6x(x + 3) - 1$

In exercises 35 through 62 you are to solve the given literal equation or formula for the specified literal symbol.

35. $y = 3x - b$ for $x$

36. $y = -ax + 7$ for $x$

37. $y - 7bx = -4k$ for $x$

38. $y - 12cz - 12 = 6$ for $z$

39. $2ax - 7by + 14 = 12$ for $y$

40. $3cx - 4by - 7 = -5$ for $y$

41.  $-5bx - y = 3x + 6y$ for x

42.  $-2ky + 2x = -3y - x + 7$ for y

43.  $-7(-2b - 3x) = 3y - 4b$ for x

44.  $-5a(-4 + y) = 3x - 16a$ for y

45.  $-0.7ax - 0.3y = -0.4a(0.5x + 0.4y)$ for y

46.  $4.3x - 4.1(1.2x - 3.5y) = 7.4x$ for y

47.  $A = bh$ for h

48.  $P = 2L + 2W$ for W

49.  $P = p + prt$ for t

50.  $P = C(1 + m)$ for C

51.  $T = 3(c - b)$ for c

52. $F_1d_1 = F_2d_2$ for $F_2$

53. $P = BR$ for $B$

54. $A = \frac{1}{2}h(b_1 + b_2)$ for $h$

55. $P = mgh$ for $m$

56. $A = p + prt$ for $p$

57. $V_1 = \frac{n_1}{n_2}V_2$ for $n_2$

58. $A = \pi r^2$ for $r$

59. $ml = \frac{yd}{R}$ for $R$

60. $g = \frac{2S}{t^2}$ for $t$

61. $Q = \frac{kA(T_2 - T_1)}{L}$ for $T_2$

62. $F = \frac{k(M_2 - M_1)}{4d^2}$ for $d$

In exercises 63 to 81 you are to solve the word problem using algebra. <u>Do not use trial and error.</u>

63. Three numbers add to 119. If the largest number is 6 more than seven times the smallest number and 10 more then twice the middle number, find the three numbers.

64. A woman is seven years younger than her husband and 1 year less than three times as old as her son. If the sum of all their ages is 89, find each of their ages.

65. A company has two divisions. Division A has now been doing business 3 times as long as division B. In 5 years division A will have been in business twice as long as division B. How long has each division been in business?

66. A company sponsored a little league baseball team by purchasing their T-shirts, pants, and hats. Each pair of pants cost $4.20 more than a shirt and $10.10 more than a hat. If the sponsor spent $726.00 on 20 complete outfits, find the cost for each shirt, each pair of pants, and each hat.

67. A cellular telephone company charges $32.25 per month plus 25¢ per minute for phone service for each of your company's mobile phones. If your company has 11 mobile phones and the charge for one month's service was $2,829.75, how many minutes of calls were made that month?

68. A painting contractor charges $15.00 per hour for his time and $12.50 per hour for each of his two assistants' time. The contractor began a job at 7 a.m. and was joined by his helpers at 8 a.m. If the three of them finished the job together and thetotal labor charge was $315.00, how many hours did each of the assistants work?

69. A propane tank's percent gauge read 13% before filling and 82% after 345 gallons of propane were added. What is the capacity of the tank?

70. A flour hopper at a commercial bakery is 1/5 full. Then 372 lb of flour is added and the hopper is found to be 4/5 full. Find the capacity of the hopper.

71. The sum of three consecutive integers is 171. Find the integers.

72. A 26 foot long counter comes in two sections. If the one section is 8 feet longer than the other section, find the lengths of the two sections.

73. Two jets originally 1,800 miles apart are flying toward each other at average speeds of 375 mi/hr and 525 mi/hr respectively. If they maintain their average speeds, how long will it take them to cover the distance?

74. How long would it take a vehicle moving at 55 mi/hr to overtake another vehicle moving at 45 mi/hr if the slower vehicle had a 2 hour head start?

75. A businesswoman drove from her office to a customer's plant and back in 18 hours (driving time). On the way back from the plant it rained and her average speed was 10 mi/hr less than her average speed going. If her average speed going was 50 mi/hr, find the round trip distance.

76. Two cars begin a trip west on Route 70 at the same time. If one travels at 9/11 the speed of the other and they are 30 miles apart after three hours, find the speed of each car.

77. Three cereal packaging machines have been scheduled for a production run of boxes of cereal. The production rates for the three machines are 1,000 boxes/hr, 1,300 boxes/hr, and 1,200 boxes/hr. If the slowest machine is started at 7 a.m. and the other two machines are started at 8 a. m., how long will it take to package a run 376,000 boxes?

78. The data processing manager of your hotel chain ordered two different grades of 8.5 in by 11 in paper in reams from your office supplier. If the higher grade paper cost $1.23 more per ream and the manager spent $2,119.00 for 250 reams of each grade, find the cost for a ream of each grade of paper.

79. How many liters of pure water must be added to 3 L of 20% alcohol in water to make 12% alcohol in water?

80. How many pounds of a coffee beans that sell for $6.00 per pound should be mixed with 10 pounds of beans that sell for $3.50 per pound to produce a bean mix that sells for $4.50 per pound?

81. An investment broker invested $5,000.00 for you in two accounts. One account earned 8% while the other lost 3%. If the combined earnings were $180.00, how much was invested in each account?

In exercises 82 to 91 you are to use proportions to solve each problem.

82. If 48.5 gal of water flow through a pipe in 2 sec, how much will flow through the pipe in 15 sec?

83. If a nursing assistant changes 7 beds in 25 minutes, how many beds will he change in 10 min if he works continuously?

84. If a 3 kg mass stretches a certain spring by 5 cm, how much will a 12 kg mass stretch the same spring?

85. If a 4 degree Celsius decrease in temperature causes a certain metal rod's length to decrease by 0.38 mm, by how much would the rod's length decrease for a 30 degree C change?

86. If a watch is gaining 2 min each day, how many minutes will it gain in 40 hr?

87. A certain car manufacturer sells both cars and trucks in the ratio of 2 to 7. If last year's sales of both cars and trucks combined was 1,892,304 units, how many of each type of vehicle were sold?

88. A wheelchair access ramp has a rise to run ratio of 1 to 4. By how many feet does the ramp rise over a 14.5 ft run?

89. A nurse has two life insurance policies. If the ratio of the value of the policies is 2:9 and the value of both policies together is $165,000.00, find the value of each policy.

90. On a drawing the drafter specified the scale as 2 in to 8 ft. Find the length of a floor joist with overlap if it is 3.625 in long on the drawing.

91. A company has $65,000.00 in one account and $15,000 in another account. What amount should be added to each account ( same for each ) so that the ratio of the two accounts is 3:2?

# CHAPTER 4
## INTRODUCTION TO FUNCTIONS AND GRAPHS

INTRODUCTION:

One of the fundamental tasks of mathematicians, physical scientists, and engineers is finding relationships between two or more variables. In this chapter we will learn about one of the most important of all mathematical relationships... the function. In addition, we will learn how to interpolate, how to draw graphs, how to interpret graphs, and how to write equations when we are given a graph or part of a graph.

OBJECTIVES:

Upon completion of this chapter you will be able to:
1. Explain what is meant by each of the terms printed in *italicized* type in each chapter section.
2. Graph first degree equations.
3. Write the equations of straight lines.
4. Solve interpolation problems.
5. Define and evaluate functions.
6. Draw and interpret non-Cartesian graphs and charts.


## 4.1 The Cartesian Coordinate System

If we place two real number lines so that we have a cross that intersects at zero on each number line, we have what is called the *Cartesian coordinate system*. See Figure 1.

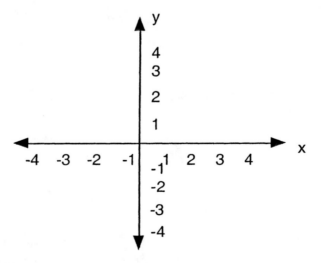

Figure 1

228

The vertical number line is called the *y-axis* while the horizontal number line is called the *x-axis*. On the y-axis, up is positive and down is negative. On the x-axis, right is positive and left is negative. The point where the two number lines intersect is call the *origin*. Invented many centuries ago by *ReneDescartes*, the coordinate system may be thought of as a naming and locating system for every point in the plane. Descartes associated what he called an *orderedpair* with every point in the plane. For example, (2, 3) is the point located two units to the right of the origin and three units above the origin. The point (-3, -2) is the point three units left of the origin and two units below the origin. See Figure 2.

Descartes called the numbers within the ordered pair the *coordinates* of the point. You should notice that the horizontal coordinate is always listed first and that the vertical coordinate is always listed second... this is why Descartes called such pairs ordered pairs.

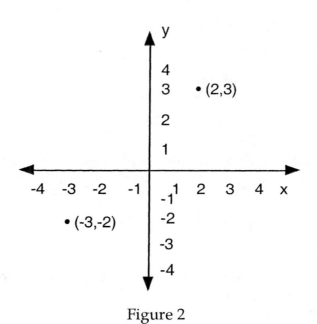

Figure 2

## Other Terminology Associated With The Cartesian Coordinate System

Descartes called the horizontal coordinate the *x-coordinate* or the *absissca*. He called the vertical coordinate the *y-coordinate* or the *ordinate*. You should note that the axes divide the plane into four regions. Descartes called these regions *quadrants* and numbered them as shown in Figure 3.

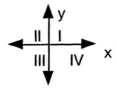

Figure 3

Please see Figure 4. It shows the signs of the ordered pairs for each quadrant. Please note that any point in Quadrant II has the ( - , +) sign pattern because it is left of and above the origin. You should study the pattern until it makes sense to you.

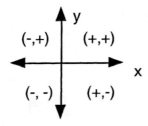

Figure 4

Two additional bits of information... many people refer to the Cartesian coordinate system as the x-y coordinate system and many others refer to it as the rectangular coordinate system.

# Exercises 4.1

For exercises 1 through 12 you are to draw a single Cartesian coordinate axes and plot all twelve points on that set of axes.

1. (3, 4)     2. (-2, 5)     3. ( -5, 0)     4. (0, 2)     5. (2, 0)     6. (0, -5 )
7. (0, 0)     8. (-3, -3)     9. ( 2, -1)     10. (-1, -2)     11. (3, -1)     12. (3, 6 )

For exercises 13 through 16 you are to tell what quadrant the point is located in without plotting the point.

13. (3, 2)          14. (-2, 1)          15. ( -2, -1)          16. (7, -2)

In exercises 17 through 23 you are to answer the question posed.

17. What is the y-coordinate of each point on the x-axis?

18. What is the abscissa of each point on the y-axis?

19. In what quadrant does the point (a,b) lie if a is positive and b is negative?

20. In what quadrant does the point (a,b) lie if a < 0 and b > 0 ?

21. In what quadrant does the point (a,b) lie if a < 0 and b < 0 ?

22. In what quadrant is the ratio of the abscissa to the ordinate positive?

23. In what quadrant is the ratio of the abscissa to the ordinate negative?

## 4.2 Graphing on the Cartesian Coordinate System

Rene Descartes' invention of the rectangular coordinate system is considered so important that many mathematicians refer to him as the father of analytic geometry. This statement prompts two questions: Why is the rectangular coordinate system so important? and, what is analytic geometry? We shall answer the latter question first. Analytic geometry is the marriage of algebra and geometry, that is, the application of algebraic methods to solving geometry problems. Now, why is the rectangular coordinate system so important? Put simply, the rectangular coordinate system allows us to draw graphs! The invention of graphing allowed mathematicians, scientists, engineers, and others to create pictures portraying the behavior of all sorts of phenomena. For example, Figure 1 is a graph showing a curve that is typical of population growth. The graph in Figure 2 shows a typical cooling curve for a high-temperature manufacturing process. The graph in Figure 3 shows the behavior that is typical of a vibrating system (not much different than the pattern seen in electrocardiograms of the human heartbeat).

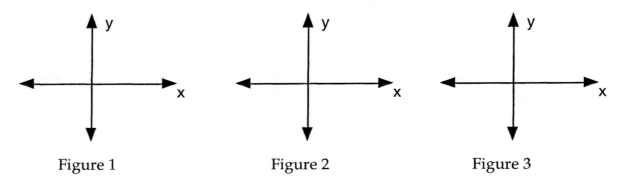

Figure 1                    Figure 2                    Figure 3

How do you draw a graph? There are many methods used for graphing. We shall begin with the most basic of these and take up another later in this chapter.

### The Point Plotting Method of Graphing

The point plotting method is just as it sounds... drawing a graph by plotting points and then connecting the dots. But, where do we get the points? The answer is that we create them using what is called an *xy table* or *function table* .

**Example 1**
Graph $y = 2x+3$.

> To create an xy table we choose x values (almost any will do) and calculate the corresponding y values by substituting into and evaluating the given equation.

| x  | y = 2x + 3 |
|----|------------|
| -1 | 2(-1)+ 3 = -2 + 3 = 1 |
| 0  | 2(0) + 3 = 0 + 3 = 3 |
| 1  | 2(1) + 3 = 2 + 3 = 5 |

so we have the ordered pair (-1,1)
so we have the ordered pair (0, 3)
so we have the ordered pair (1, 5)

Now we simply plot these points and draw a line through them. See Figure 4.

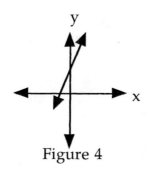

Figure 4

## Example 2

Graph the equation y = 4 - 2x.

Making our xy table:

| x  | y = 4 - 2x |
|----|------------|
| -1 | 4 - 2(-1) = 4 + 2 = 6 |
| 0  | 4 - 2(0) =  4 + 0 = 4 |
| 1  | 4 - 2(1) =  4 - 2 = 2 |

The graph of y = 4 - 2x is shown in Figure 5.

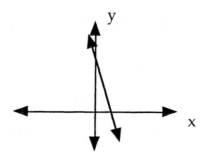

Figure 5

Please note that the graph of y = 4 - 2x is a straight line. All equations containing exclusively x and y, each to the first power, will have straight-line graphs. That is why such equations are often referred to as *linearequations*.

Many times the equation will be presented to us in a form that requires us to solve for y before we can make our xy table. Example 3 demonstrates such a situation.

## Example 3
Graph the equation 3y - 6x = 9.

We must solve for y first. Adding 6x to both sides of the equation we have:

$$3y = 9 + 6x$$

Dividing both sides of the equation by 3 yields:

$$y = 3 + 2x \text{ or } y = 2x + 3$$

Making our xy table:

| x | y = 3 + 2x |
|---|---|
| -1 | 3 + 2(-1) = 3 - 2 = 1 |
| 0 | 3 + 2(0) = 3 + 0 = 3 |
| 1 | 3 + 2(1) = 3 + 2 = 5 |

The graph of y = 3 + 2x is shown in Figure 6.

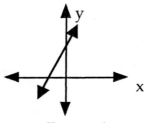

Figure 6

Sometimes when we solve an equation for y the result will have a fraction. Example 4 demonstrates such a situation.

## Example 4
Graph the linear equation 2y - 3x = 4.

We must solve for y first. Adding 3x to both sides of the equation we have:

$$2y = 4 + 3x$$

Dividing both sides of the equation by 2 yields:

$$y = (4 + 3x)/2$$

When we select x values for our xy table we want to avoid values that will result in fractions, since fractional values are difficult to graph. For example,

234

if we choose x = 1, then y = ( 4 + 3(1))/2 = ( 4 + 3)/2 = 7/2. So let's choose x values that will result in y values that are not fractions. For example, choosing x = 0 yields y = ( 4 + 3(0))/2 = 4/2 = 2. Now that we have one x value that works, all we have to do is add or subtract the denominator (in this case 2) to that x value and we will have two more x values that work for us. So 0 + 2 = 2 will work and so will 0 - 2 = -2.

Making our xy table:

| x | y = (4 + 3x)/2 |
|---|---|
| -2 | (4 + 3(-2))/2 = (4 - 6)/2 = -2/2 = -1 |
| 0 | ( 4 + 3(0))/2 = 4/2 = 2 |
| 2 | (4 + 3(2))/2 = (4 + 6)/2 = 10/2 = 5 |

The graph of y = (4 + 3x)/2 is shown in Figure 7.

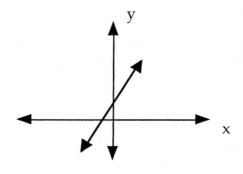

Figure 7

## Graphing Horizontal or Vertical Lines

An equation of the form y = k (where k is just some number) will have a horizontal line for its graph.

**Example 5**
Graph the equation y = 3.

The graph of the equation y = 3 is simply a horizontal line three units above the x-axis. Why? Well, every point on that line has y coordinate 3 and so we call the line y = 3. See Figure 8.

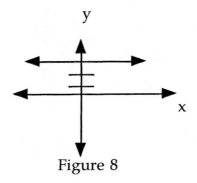

y

x

Figure 8

An equation of the form x = k (where k is just some number) will have a vertical line for its graph.

## Example 6
Graph the equation x = -2.

The graph of the equation x = -2 is simply a vertical line two units left of the y-axis. Why? Well, every point on that line has x coordinate -2 and so we call the line x = -2. See Figure 9.

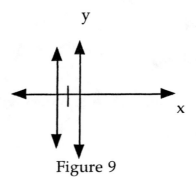

y

x

Figure 9

# Exercises 4.2

You are to graph each of the following equations on graph paper. For sloped (nonhorizontal-nonvertical) lines you must use three ordered pairs in your xy table. The answers may be found in the rear of the book.

1. $y = 4$

2. $x = 3$

3. $y + 6 = 8$

4. $y = x + 2$

5. $y = 2 - x$

6. $y = x$

7. $y = -x$

8. $y = 3x - 7$

9. $3x + y = 5$

10. $y - 2x = 1$

11. $3x + 4y = 7$

12. $x + 5y = 19$

13. $x + y = 4$

14. $5x - 2y = 17$

15. $4x - 5y = 17$

16. $2.5 = x$

17. $y = -5/2$

18. $x - 5 - y = 0$

19. $3x = -3y$

20. $-x = 3$

## Graphing Applications

21. A budget line shows all of the different combinations of goods a person can buy with a given budget, B. If a given company has \$400 to spend on 2 goods, "a" and "b" and a = \$20 and b = \$10, graph the budget line using the equation $ax + by = B$

22.   A budget line shows all of the different combinations of goods a person can buy with a given budget, B. If a given company has $200 to spend on 2 goods, "a" and "b" and a = $30 and b = $20, graph the budget line using the equation ax + by = B

23.   Straight line depreciation is a linear equation. The current value, y, of a certain company's piece of equipment after x years is y = 20,000 − 2000x.

   a)   Find the initial value of the equipment.

   b)   Find the value after 3 years.

   c)   Find the value after 5 years.

   d)   Graph the depreciation line.

## 4.3 Writing Linear Equations

In section 4.2 our task was this: Given a linear equation, we had to draw its graph. In this section our task is this: Given a straight line graph or part of that graph, write the linear equation.

Before we can learn to write linear equations we must understand the concept of slope.

## The Slope of a Straight Line Graph

The slope of a line is the ratio of the rise (vertical change) to the run (horizontal change). Generally symbolized by the letter m, the slope is:

$$m = rise/run$$

Let's take any line and locate two points $(x_1, y_1)$ and $(x_2, y_2)$ on that line. See Figure 1. Please note that the rise shown is the difference of the y coordinates $y_2 - y_1$. Please also note that run is the difference of the x coordinates or $x_2 - x_1$. And so the slope is

$$m = \frac{y_2 - y_1}{x_2 - x_1}$$

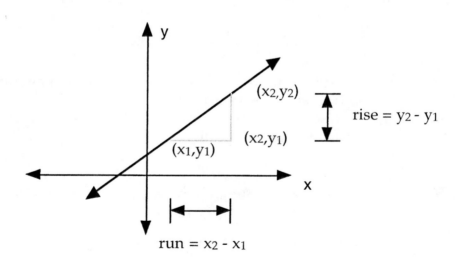

Figure 1

239

## Example 1
Find the slope of the line that passes through the points (1, 3) and (3, 7).

$$m = \frac{7-3}{3-1} = \frac{4}{2} = 2,$$

which means the graph is rising two units for every 1 unit it moves right.

You should note that it does not matter which point we let be $(x_1, y_1)$ and which we let be $(x_2, y_2)$. We could have let (1, 3) be $(x_2, y_2)$ instead of (3, 7) as we did above. This would yield the same answer. To prove this to you:

$$m = \frac{3-7}{1-3} = \frac{-4}{-2} = 2$$

## Example 2
Find the slope of the line that passes through the points (-2, 4) and (5, 2).

$$m = \frac{2-4}{5-(-2)} = \frac{-2}{7}$$

A slope of -2/7 means that the graph is moving down two units for every seven units it moves right.

## Example 3
Find the slope of the line that passes through the points (-2, 1) and (2, 1).

$$m = \frac{1-1}{2-(-2)} = \frac{0}{4} = 0$$

A slope of 0 means that the graph is not moving up or down as it moves right. **Therefore, the line must be a horizontal line**. Draw the two points on a Cartesian coordinate system and then draw a line through them and you will see that the line is indeed a horizontal line.

## Example 4
Find the slope of the line that passes through the points (-1, 2) and (-1, 4).

$$m = \frac{4-2}{-1-(-1)} = \frac{2}{0}$$ is undefined since we cannot divide by zero.

An undefined slope means that the graph is moving straight up. **Therefore, the line must be a vertical line**. Draw the two points on a Cartesian coordinate system and then draw a line through them and you will see that the line is indeed a vertical line.

## The Point Slope Formula for the Equation of a Straight Line

If we take the formula for slope and let the point ($x_2$, $y_2$) be free to be any point (x y) on the line (except ($x_1$, $y_1$)), then we have:

$$m = \frac{y - y_1}{x - x_1}$$

If we multiply both sides of the equation by x - $x_1$ we have:

$$(x - x_1)m = y - y_1$$

Rearranging we have:

$$y - y_1 = m(x - x_1)$$

This last equation is called the point-slope formula for writing the equation of a straight line. It is called the point-slope formula because you need the slope m and a point ($x_1$, $y_1$) to be able to use it.

### Example 5
Find the equation of the line that passes through the point (1, 2) and which has slope 3.

Substituting into the point-slope formula we have:

$$y - 2 = 3(x - 1)$$

Simplifying we have:

$$y - 2 = 3x - 3$$
$$\text{or} \quad y = 3x - 1$$

### Example 6
Find the equation of the line that passes through the points (1, 2) and (5, 4).

Here we are missing the slope m, but we can calculate it since we have two points.

$$m = \frac{4 - 2}{5 - 1} = \frac{2}{4} = \frac{1}{2}$$

Then using either point and substituting into the point-slope formula we have:

$$y - 2 = \frac{1}{2}(x - 1)$$

Simplifying we have:

$$2y - 4 = x - 1$$

241

$$\text{or } 2y = x + 3$$

$$\text{or } y = \frac{x+3}{2}$$

Could we have used the other point (5, 4) ? The answer is yes. You should substitute m = 1/2 and the point (5, 4) into the point slope equation and prove that the same equation will result.

## The Intercepts of a Straight Line Graph

The *intercepts* of any graph are the points where the graph crosses the axes. See Figure 2. Please note that at the point where the graph crosses the y-axis, its x-coordinate must be zero. Please also note that at the point where the graph crosses the x-axis, its y-coordinate must be zero. We can use these two geometric facts to find the intercepts of any graph.

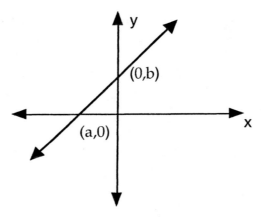

Figure 2

**Example 7**
Find the intercepts of 2x + 3y = 6.

To find the x-intercept we set y equal to 0 and solve for x:

$$2x + 3(0) = 6$$
$$2x = 6$$
$$x = 3$$

And so our x-intercept is the point (3, 0).

To find the y-intercept we set x equal to 0 and solve for y:

$$2(0) + 3y = 6$$
$$3y = 6$$
$$y = 2$$

242

And so our y-intercept is the point (0, 2).

Not only have we found the intercepts, but we have enough information to actually graph the line (See Figure 3). *This provides us with a very easy method for graphing* .

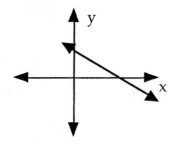

Figure 3

Now that we have learned about intercepts we are ready for the following formula.

## The Slope-Intercept Formula for the Equation of a Straight Line

$$y = mx + b$$

This formula is called the slope-intercept formula because we need to have both the the slope (m) and the y-coordinate (b) of the y-intercept to be able to use it.

### Example 8
Write the equation of the line with slope 3 which passes through the point (0,-4).

If we realize that (0, -4) is the y-intercept, then all we need do is this:

$$y = 3x + (-4) \text{ or } y = 3x - 4$$

# Exercises 4.3

In exercises 1 through 9 you are to find the slope of the line passing through each of the following pairs of points. Show all work in the spaces provided.  The answers may be found in the rear of the book.

1. (5, 2) and (3 ,3)           2. (-4, 3) and (2, 3)           3. (3, -1) and (0, 3)

4.  (5, 6) and (3, 4)          5.  (6, -7) and (6, 1)          6.  (0, 7) and (0, 8)

7.  (2, 0) and (3, 0)          8.  (-10, 4) and (2, -5)          9.  (5, 7) and (-6, -3)

In exercises 10 through 15 you are to find the equation of the line passing through the given point and having the given slope.

10.  (1, 3) and m = 4          11.  (0, -1) and m = -2          12.  (2, 7) and m = 3/5

13.  (2, 3) and m = 0          14.  (-3, -1) and m = -5/8          15.  (1, -8) and m is
                                                                        undefined

In exercises 16 through 24 you are to find the equation of the line passing through the given pair of points.

16.  (5, 2) and (3, 3)          17.  (-4, 3) and (2, 3)          18.  (3, -1) and (0, 3)

19.  (5, 6) and (3, 4)          20.  (6, -7) and (6, 1)          21.  (0, 7) and (0, 8)

244

22.  (2, 0) and (3, 0)                23.  (-10, 4) and (2, -5)                24.  (5, 7) and (-6, -3)

In exercises 25 through 28 you are to find the x- and y-intercepts of the line represented by the given equation.

25.  $3x - 2y = 6$          26.  $x + 2y = 3$          27.  $3x + 4y = 12$          28.  $-2x + y = -4$

## 4.4  Linear Interpolation

*Interpolation* literally means to find a value **in between** two other values. Consider the pressure-volume table shown below.

| Pressure  (psi or lb/in²) | Volume  (ci or in³) |
|:---:|:---:|
| 20 | 100 |
| 25 | 90 |
| 30 | 80 |

Now suppose we wish to know what pressure is associated with a volume of 96 in³.

Method 1:  Classic Interpolation Analysis

| 20 | 100 | Insert the new values in as shown. |
|:---:|:---:|:---|
| x | 96 | Then write a proportion based on distance. |
| 25 | 90 | |

$$\frac{x-20}{25-20}=\frac{96-100}{90-100}=\frac{\text{distance from x to 20}}{\text{distance from 25 to 20}}=\frac{\text{distance from 96 to 100}}{\text{distance from 90 to 100}}$$

Simplifying:

$$\frac{x-20}{5}=\frac{-4}{-10} \quad \text{or} \quad \frac{x-20}{5}=\frac{2}{5} \quad \text{or} \quad x-20=2 \text{ or } x=22$$

Method 2. Writing a linear equation.

Using (100, 20) and (90, 25) as  (v, p) points, the slope is then:

$$m=\frac{25-20}{90-100}=\frac{5}{-10}=-\frac{1}{2}$$

Then our equation is:

$$p-20=\frac{-1}{2}(v-100) \text{ or } p-20=\frac{-v}{2}+50 \text{ or } p=\frac{-v}{2}+70.$$

We now have a formula that will allow us to calculate the pressure associated with any volume. Now substituting v = 96 into our equation we have:

$$p = -96/2 + 70 = -48 + 70 = 22 \text{ psi.}$$

There are two **major benefits** to this second method:
   1. If you need to calculate many values it is easier to use the formula than to repeatedly set up the classical interpolation proportion, and

2. We could solve $p = -v/2 + 70$ for v and have a formula for calculating the volume if the pressure is known.

# Exercises 4.4

In exercises 1 through 6 you are to use Table 1 and the classical interpolation (proportion) method to find the requested value. Show all work in the spaces provided. The answers may be found in the rear of the book.

**Table 1**
**Monthly Life Insurance Premiums Per $10,000 Coverage For Female Nonsmokers**

| Age | Monthly Life Insurance Premium (per $10,000 coverage) |
|---|---|
| 20 | $2.00 |
| 30 | $3.16 |
| 40 | $4.32 |
| 50 | $5.48 |
| 60 | $6.64 |

1. Find the monthly premium for a woman age 55 with $10,000 coverage.

2. Find the monthly premium for a woman age 28 with $10,000 coverage.

3. Find the monthly premium for a woman age 36 with $75,000 coverage.

4. If the premium is $6.90 **for $20,000 coverage,** find the woman's age.

5. If the premium is $24.50 **for $50,000 coverage,** find the woman's age.

6. If the premium is $15.57 **for $30,000 coverage,** find the woman's age.

In exercises 7 through 12 you are to use Table 2.

**Table 2**

**Average Weekly Income Based on Age in Pinesville Ohio**
(assuming a linear relationship)

| Age (years) | Income (dollars) |
|---|---|
| 20 | 300 |
| 30 | 450 |
| 40 | 600 |
| 50 | 750 |

7.  Write an equation relating age (a) and income (i).  Hint: Treat a as x in the slope and point slope formulas.

8.  Find the income associated with an age of 25 years.

9.    Find the income associated with an age of 44 years.

10.    Solve your equation from problem 7 for the variable a.

11.    Find the age associated with an income of $570.00.

12.    Find the age associated with an income of $330.00.

In exercises 13 through 16 you are to use Table 3 and either interpolation method.

**Table 3**
**Height Weight Relationship for Adolescent Males**

| Height (inches) | Weight (pounds) |
|-----------------|-----------------|
| 24              | 51              |
| 36              | 80              |
| 48              | 109             |
| 60              | 137             |

13. Find the recommended weight for a boy who is 30 inches tall.

14. Find the ideal weight for a boy who is 50 inches tall.

15. If a boy weighs 105 lb, what should his height be?

16. If a boy weighs 120 lb and is 50 inches tall, is he too heavy for his height? If so, by how many pounds?

## 4.5 Functions

One of the fundamental tasks of mathematicians, physical scientists, and engineers is finding relationships between two or more variables. In this chapter section we will learn about one of the most important of all mathematical relationships... the function.

Before we define functions let's look at several examples that demonstrate common usage of the word function.

### Example 1
Suppose your pay rate is $9.25 per hour. Then your weekly pay P (without overtime) is:

$$P = 9.25t, \text{ where t is the number of hours worked.}$$

Here we say that your pay P is a **function** of the time t since the amount you will be paid **depends** upon the amount of time t you worked that week.

### Example 2
Suppose a cab ride costs $2.00 plus 50¢ per mile. Then the cost C of a ride is:

$$C = 200 + 50m \text{ where m is miles (in cents)}$$
OR
$$C = 2.00 + 0.50m \quad \text{(in dollars)}$$

We say that the cost of the cab ride C is a **function** of the mileage m since the cost **depends** upon the number of miles that you ride.

## Terminology

In $P = 9.25t$, P is called the dependent variable (since pay depends on the time worked) and t is the independent variable. In $C = 2.00 + 0.50m$, C is called the dependent variable (since the cost depends on the mileage driven) and m is the independent variable.

Now we are ready to define the term function.

A *function* is a relationship between two variables such that for every value of the independent variable there is exactly one (one and only one) value of the dependent variable.
What is this restriction that says there can be one and only one value of the dependent variable for each value of the independent variable?
Consider Example 1 above. If you worked 10 hours, then your pay would be

P = 9.25(10) = 92.50. So $92.50 is the one and only one dependent variable value associated with the independent variable value 10.

Consider Example 2 above. If you rode 10 miles, then the cost of your cab ride would be C = 2.00 + 0.50(10) = 2.00 + 5.00 = 7.00. So $7.00 is the one and only one dependent variable value associated with the independent variable value 10.

Is it ever possible to have a formula that yields two values when only one value is substituted in? Yes. Here is an example:

## Example 3
Give an example of a relationship that is not a function.

Consider $y = x^2$. It is a function of x since each value of x yields only one value of y. But, if we solve for x we get:
$$x = \pm\sqrt{y}$$
which is not a function of y, since each value of y yields two values (plus or minus) of x.

## Notation

The notation $y = f(x)$, read "y equals f of x," says that y is a function of x and that y is the dependent variable and x is the independent variable.

## Example 4
What does the notation $t = f(s)$ mean?

The notation $t = f(s)$, read "t equals f of s," says that t is a function of s and that t is the dependent variable and s is the independent variable.

## Example 5
What does the notation $w = h(r)$ mean?

The notation $w = h(r)$, read "w equals h of r," says that w is a function of r and that w is the dependent variable and r is the independent variable.

## Evaluating Functions

We evaluate functions just as we did formulas in earlier sections.

252

**Example 6**
Given $y = f(x) = 2x + 3$, evaluate f given $x = 3$.

This means substitute 3 in wherever there is an x and then evaluate
$$y = f(3) = 2(3) + 3 = 6 + 3 = 9.$$

**Example 7**
Given $y = h(x) = 5x^2 + 2x - 4$, evaluate h given $x = -2$.

This means substitute -2 in wherever there is an x and then evaluate:
$$y = h(-2) = 5(-2)^2 + 2(-2) - 4 = 5(4) + 2(-2) - 4 = 20 + (-4) - 4 = 12.$$

**Example 8**
Given $y = f(x) = 4x + 3$ evaluate f given $x = 2x - 5$.

This means substitute 2x - 5 in wherever there is an x and then evaluate:
$$y = f(2x - 5) = 4(2x - 5) + 3 = 8x - 20 + 3 = 8x - 17.$$

**Example 9**
Given $y = f(x) = 3x + 7$ and $y = g(x) = 4 - 5x$ find $f(g(x))$.

This means substitute g(x) in wherever there is an x and then evaluate:
$$y = f(g(x)) = 3(4 - 5x) + 7 = 12 - 15x + 7 = 19 - 15x.$$

# Exercises 4.5

In exercises 1- 10 you are to consider the given function and identify both the dependent variable and independent variable.

1.  A car rental costs $10.00 per day plus 18 cents per mile, so its cost per day is $C = 10.00 + 0.18m$, where m is the number of miles driven.

2.  An electrician charges $25.00 plus $30.00 per hour for a service call. So the cost of a service call is $C = 25.00 + 30.00h$, where h is the number of hours worked.

3.  AT&T charges $4.95 per month plus 10¢ per minute for long distance calls in the continental U.S. So the per month cost of long distance service is $C = 4.95 + 0.10m$, where m is the number of minutes of calls.

4. A glass company charges a $400.00 set-up fee and $4.75 per lens to make a headlight lens for an auto manufacturer. So the cost of one run of these lenses is C = 400.00 + 4.75n, where n is the number of lenses produced.

5. $y = f(x) = 3x - 5$

6. $P = h(r) = 12 - 4r + r^2$

7. $q = f(t) = t^3/4 + t^2/3 + 6$

8. $y = f(x) = (2x - 3)^2$

9. $w = k(z) = z^3 - 4z^2 + z + 11$

10. $z = h(x) = 4x^3 + 2x^2 + 3x + 5$

In exercises 11 to 20 you are to evaluate the specified function at the independent variable value provided.

11. Evaluate the function in problem 1 if the mileage driven is 240 miles.

12. Evaluate the function in problem 2 if the electrician works 14.5 hours.

13. Evaluate the function in problem 3 if 112 minutes of calls were made.

14. Evaluate the function in problem 4 if 5,000 lenses were manufactured.

15. Evaluate the function in problem 5 if x = - 1.

16. Evaluate the function in problem 6 if r = - 2.

17. Evaluate the function in problem 7 if t = 4.

18. Evaluate the function in problem 8 if x = 3.

19. Evaluate the function in problem 9 if z = - 1.

20. Evaluate the function in problem 5 if x = 5.

## 4.6  Non-Cartesian Graphs and Charts

In this section we learn about bar graphs, broken line graphs, histograms, and pie charts. We will begin with *pie charts*.

### Example 1
A certain police department has 20 officers assigned to the first precinct, 16 assigned to the second precinct, 12 assigned to the third precinct, 14 to the fourth, and 32 assigned to other areas. Draw a pie chart describing this situation.

First we find the total number of officers. 20 + 16 + 12 + 14 + 32 = 94 officers.

A circle has 360°. So we multiply 360° by the fraction formed by dividing the part by the total.

$$\frac{20 \text{ officers}}{94 \text{ officers}} \cdot 360 \text{ degrees } = 77 \text{ degrees}$$

$$\frac{16 \text{ officers}}{94 \text{ officers}} \cdot 360 \text{ degrees } = 61 \text{ degrees}$$

$$\frac{12 \text{ officers}}{94 \text{ officers}} \cdot 360 \text{ degrees } = 46 \text{ degrees}$$

$$\frac{14 \text{ officers}}{94 \text{ officers}} \cdot 360 \text{ degrees } = 54 \text{ degrees}$$

$$\frac{32 \text{ officers}}{94 \text{ officers}} \cdot 360 \text{ degrees } = 123 \text{ degrees}$$

We divide the circle into sections accordingly and then label the sections:

Police Officer Assignments by Precinct

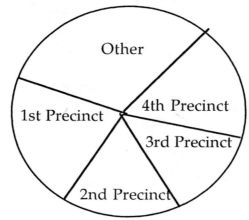

255

Now let's tackle a *broken line graph* and a *bar graph*.

## Example 2

During a five day period, a hotel registered the number of guest listed in the table below. Show the results in a broken line graph.

| Day | Number of Guests |
|-----|------------------|
| Monday | 45 |
| Tuesday | 63 |
| Wednesday | 70 |
| Thursday | 66 |
| Friday | 42 |

Guest Registration By Weekday

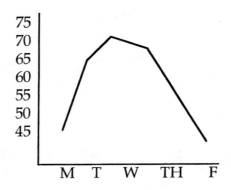

## Example 3

The average price of a three bedroom house in five different cities is given in the table below. Draw a bar graph showing these results.

| Price (in thousands of $) | City |
|---------------------------|------|
| 103 | Columbus |
| 114 | Cincinnati |
| 89 | Dayton |
| 96 | Toledo |
| 125 | Cleveland |

256

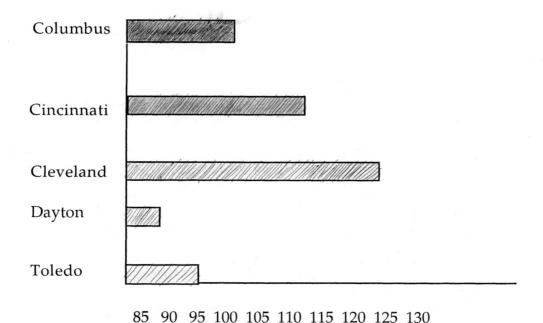

Average Home Prices in Five Ohio Cities

Columbus

Cincinnati

Cleveland

Dayton

Toledo

85  90  95  100  105  110  115  120  125  130

We could have made a vertical bar graph rather than a horizontal bar graph. That is, we could have put the prices on the vertical axis and the cities on the horizontal axis, and then our bars would have run vertically. Either way is fine!

Histograms are different. The width of each bar corresponds to a range of numbers called the *class interval*.The height of each bar corresponds to the number of occurrences in each class and is called the *class frequency* . All class intervals must have equal widths and there cannot be any gaps between the class boundaries... and thus no gaps between the bars on the histogram itself.

## Example 4

United Parcel Service (UPS) conducted a study of heavier packages delivered by one driver during a one week long period. The following table shows the weights and number of packages for parcels that weighed 18 to 35 pounds. Show the results in a histogram.

| Weights (pounds) | Number of Packages |
|---|---|
| 18-20 | 12 |
| 21-23 | 19 |
| 24-26 | 24 |
| 27-29 | 17 |
| 30-32 | 15 |
| 33-35 | 5 |

The class intervals (horizontal axis) are the weights and the class frequencies (vertical axis) are the numbers of packages.

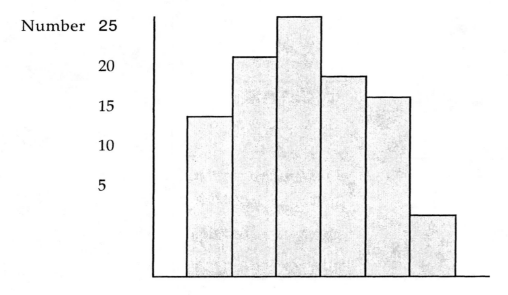

Weights (pounds)

Weights of Packages

# Exercises 4.6

1.  A restaurant's receipts for the first six months are shown in the table. Show the results in a broken line graph.

| Month | Value (in $1,000) |
|---|---|
| January | 75 |
| February | 92 |
| March | 106 |
| April | 110 |
| May | 101 |
| June | 83 |

2.  A caterer's gross receipts are shown in five categories. Display these receipts in a pie chart.

|  Category | Receipts |
|---|---|
| Business Luncheons | 22,500.00 |
| Club Luncheons | 12,400.00 |
| Weddings | 37,250.00 |
| Funerals | 7,400.00 |
| Miscellaneous | 9,700.00 |

3.  Consider the table shown in Exercise 1. Draw a bar graph showing the results.

4.   A corrections officer looked at the test scores of 39 juvenile offenders who are enrolled in the institution's high school program. The results are shown in the table below. Put the results into a histogram.

| Points (Out of 100) | Number of Inmates |
| --- | --- |
| 60 to 69 | 5 |
| 70 to 79 | 10 |
| 80 to 89 | 13 |
| 90 to 99 | 11 |

5.   A large hotel has 30 housekeeping and maintenance staff, 10 desk clerks, 26 restaurant employees, 4 managers, and 6 office and accounting staff. Draw a pie chart describing this situation.

6. During a six month period last year, a certain police chief reported the number of breaking and entering cases (B & E) listed in the table below. Show the results in a broken line graph.

| Month | Number of B & E's |
|-------|-------------------|
| May | 15 |
| June | 11 |
| July | 17 |
| August | 12 |
| September | 7 |
| October | 5 |

7. A restaurant chain conducted a study of meat sales during the past month. The following table shows the weights and number of restaurant units that sold that much meat. Show the results in a histogram.

| Weights (pounds) | Number of Units |
|------------------|-----------------|
| 1,500-1,999 | 7 |
| 2,000-2,499 | 21 |
| 2,500-2,999 | 29 |
| 3,000-3,499 | 19 |
| 3,500-3,999 | 15 |
| 4,000-4,499 | 11 |

8.  The average income of police officers in five different cities is shown in the table. Draw a bar graph showing these results.

| Income (in thousands of $) | City |
|---|---|
| 32.5 | Robertsburg |
| 27.2 | Glenwood |
| 39.6 | New Ritchey |
| 24.7 | Chesterwood |
| 41.3 | Kennwood |

## Exercise Answers

<u>1.1 Answers</u>

1.  A number with a unit:  5 lb, 4 in, 3 m, 6 sec, and 13 mm.

2.  To standardize units for travel, trade, and construction. Everybody uses standard units rather than local units.

3.  To eliminate the forced use of fractions or large numbers. For example: 3/4 foot is 9 inches and 13,200 ft is 2.5 miles.

4.  They know what the other is trying to communicate. For example: 3 oz means 3 oz not 3 sec or 3 m.

5.  When you weigh yourself you are comparing your body weight to a standard known as the pound.  If you weigh 120 pounds then your weight was compared to the standard and found to be 120 times as great.

6.  The customary system.

7.  a) about 1.5 mm.
    b) about 1.5 to 2 cm.
    c) about 2.5 cm.

8.  a) 2 m
    b) 4 kg
    c) 3 cm
    d) 50 lb
    e) 1 L
    f) 1 kg

9.  25,000 dollars   10. 25,000 W, yes   11. 250,000 L   12. 1/10 L

13. 1/5 A            14. 450,000 J       15. 1,200 g     16. 3,500,000 L

17. 750,000 Cal      18. 4/5 m           19. 240 g       20. 37/100 L

21. 2.5 m            22. 11/20 mL        23. 3/10 sec    24. 1/4 sec

25. 25,000 tons      26. 50,000,000 tons 27. 1,700,000 m 28. 250 m

29. 1,200,000 cd     30. 3.5 g           31. 1/100 m     32. 500,000 W

33. 500,000 g　　　34. 9,5000,000 m　　35. 2,300,000 m　　36. 35/1 m

1.2 Answers

1. $28 \text{ ft} \times \dfrac{12 \text{ in}}{1 \text{ ft}} = 336 \text{ in}$　　　　2. $70 \text{ oz} \times \dfrac{1 \text{ lb}}{16 \text{ oz}} = 4.38 \text{ lb}$

3. $30 \text{ mi} \times \dfrac{1.609 \text{ km}}{1 \text{ mi}} = 48.27 \text{ km}$　　　4. $775 \text{ mm} \times \dfrac{1 \text{ cm}}{10 \text{ mm}} = 77.5 \text{ cm}$

| | | | |
|---|---|---|---|
| 5. 226.76 kg | 6. 37.84 L | 7. 305 m | 8. 17 lb 2 oz |
| 9. 2 mi 190 ft | 10. 712.46 m | 11. 105.2 cm | 12. 4 lb |
| 13. 24 cm | 14. 1,200 μm | 15. 9.07 t | 16. 250 dam |
| 17. 372.9 mi | 18. 0.3 mg | 19. 362.88 g | 20. 12 tsp |
| 21. 20 pt | 22. 4.30 bushels | 23. 2.54 barrels | 24. 163.8 gal |
| 25. 250 mL | 26. 5 cc | 27. 450 cc | 28. 11,700 sec |
| 29. 43,200 min | 30. 3.67 hr | 31. 21,600 sec | 32. 2,520 min |
| 33. 48.82 in | 34. 119.38 mm | 35. 304.31 L | 36. 0.25 ton |
| 37. 1.83 m | 38. 0.48 gal | 39. 1.34 mi | 40. 2.65 km |
| 41. 10,884.35 kg | 42. 8.45 fl oz | 43. 1 Mton | 44. 1,584 in |
| 45. 1.13 kg | 46. 1,587.3 kg | 47. 5 Mg | 48. 1,126.3 km |
| 49. 20 lb 4 oz | 50. 3 mi 4430 ft | 51. 630.25 m | 52. 19.16 tons |
| 53. 1.81 kg | 54. 52.84 gal | 55. 0.28 in | 56. 951.12 gal |
| 57. 88.2 lb | 58. a) 20 tons  b) 18,140.59 kg | | 59. 683.83 km |
| 60. 1,930.81 cm | 61. $680 | 62. $11,625 | 63. 71.12 cm |
| 64. 23.85 m | | | |

## 1.3 Answers

1. $11.5 \text{ ft}^2 \times \dfrac{144 \text{ in}^2}{1 \text{ ft}^2} = 1,656 \text{ in}^2$

2. $175 \text{ cm}^2 \times \dfrac{100 \text{ mm}^2}{1 \text{ cm}^2} = 17,500 \text{ mm}^2$

3. $300 \text{ cm}^2 \times \dfrac{1 \text{ in}^2}{6.45 \text{ cm}^2} = 46.51 \text{ in}^2$

4. $105 \text{ m}^2 \times \dfrac{10.76 \text{ ft}^2}{1 \text{ m}^2} = 1,129.8 \text{ ft}^2$

5. $2,295 \text{ ft}^2$

6. $200 \text{ ft} \times 750 \text{ ft} = 150,000 \text{ ft}^2$ now convert to acres

$\quad 150,000 \text{ ft}^2 \times \dfrac{1 \text{ acre}}{43,560 \text{ ft}^2} = 3.44 \text{ acres}$

7. 41.14 gal

8. $3,160

9. 96 yd²

10. 67.56 yd²

11. 266.67 buckets

12. 755.04 ft

13. 4.05 ft³

14. 380 cc

15. 2.75 L

16. 5.74 L

17. 133.67 ft³

18. 16,772.25 ft³

19. 0.35 m³

20. 65.05 cc

21. 25.36 fl oz

22. 897.72 gal

23. 134.23 in³

24. a) 300 m³  b) 300 kL  c) 79,260.24 gal

25. 112.22 min

26. 7.96 loads

27. 5.53 min

28. $\dfrac{60 \text{ mi}}{\text{hr}} \times \dfrac{1.609 \text{ km}}{1 \text{ mi}} = 96.54 \text{ km} / \text{hr}$

29. 0.013 km/sec

30. 14,764.5 ft/min

31. 954.55 mi/hr

32. 11.48 km/L

33. Car B by 0.75 km/L

34. $\dfrac{32 \text{ lb}}{\text{in}^2} \times \dfrac{1 \text{ kg}}{2.205 \text{ lb}} \times \dfrac{1,550 \text{ in}^2}{1 \text{ m}^2} = 22,494.33 \text{ kg} / \text{m}^2$

35. 0.07 lb/in²

36. 30.5 min

37. 25.71 bur/week

38. 625.71 assaults/yr

39. $143.57

40. $3.5\,\text{oz} = \dfrac{540\,\text{Cal}}{2\,\text{oz}} = 945\,\text{Cal}$

41. 11,040 Cal

42. 175 min

43. 3660.8 calls/yr

44. 12.5 hr

45. 5

46. 20 hours

47. 2.63 min

48. $277,200

49. no

50. $13,584,375

51. 1,531.25

52. a) 14.99 oz   b) 9.45 in   c) 2.6 times

<u>1.4 Answers</u>

1. $\dfrac{\$15.60}{20\,\text{lb}} = \$0.78\,/\,\text{lb}$

2. $6.65/ft

3. $\dfrac{\$31.12}{25\,\text{lb}} = \$1.24\,/\,\text{lb}$         $\dfrac{\$29.77}{10\,\text{kg}} \times \dfrac{1\,\text{kg}}{2.205\,\text{lb}} = \$1.35\,/\,\text{lb}$

    The 25 lb box is the better buy.

4. US supplier

5. 25 packs

6. $0.30/lb

7. 2,000 ticket lot

8. $\dfrac{\$16.50}{1\,\text{gal}} \times \dfrac{1\,\text{gal}}{3.781\,\text{L}} = \dfrac{\$4.36}{1\,\text{L}}$

    145% of $\dfrac{\$4.36}{1\,\text{L}} = \$6.32$ per liter

9. $4.64

10. $360

11. $\dfrac{\$1.09}{1 \text{ kg}} \times \dfrac{1 \text{ kg}}{2.205 \text{ lb}} = \$0.49 / \text{lb}$   So loss per pound is $\$0.60 - \$0.49 = \$0.11$

The store sold $1,500 \text{ kg} \times \dfrac{2.205 \text{ lb}}{1 \text{ kg}} = 3,307.5 \text{ lb}$

So, $3,307.5 \times \$0.11 = \$363.83$ loss

Percent loss $= \dfrac{\$0.11}{\$0.60} \times 100\% = 18.33\%$

12. 95301.81 gal/yr,  104,831.99 gal        13.  $0.37 profit, 63.79% profit

14. $2.5 profit, 71.43% profit        15.  $2.29 profit, 134.71% profit

16. $2.40 profit, 685.71% profit        17.  local store        18.  Canadian, $22.23

19. $1,261,367,438    20.  a) 42   b) $1,292,550

## 1.5 Answers

1.    a)  numbers that include both a numerical value and a unit of measurement.
      b)  a denominate number such as 5 ft 9 in.
      c)  a specific measurement of length, mass, temperature or time.
      d)  unit that is used as a base for auxiliary units such as feet (base) and inch (auxiliary).
      e)  added to customary system (feet has inches, miles, and yards).
      f)  any unit created using multiplication, division or both of two or more base units.
      g)  a change from one unit of measurement to another.
      h)  a change within a system of measurement.

2.    Metric units of measurement including all prefixes (like centi and milli) are based on powers of ten.  Customary base units (feet) and their auxiliary units (inches) are based on multiples of 2 and 6.  Metric units are easier to use for conversions.

3.  100,000 W

4.  1/10 g

5.  2.5 ML

6.  25,000,000 dol

7.  1/10 L

8.  6/25 L

9.  50 V

10.  5 m

11.  220 m

12.  1/5 A

13.  450 kg

14.  3,500 kN

15.  Change $0.25 \text{ lb} \times \dfrac{16 \text{ oz}}{1 \text{ lb}} = 4 \text{ oz}$  So we have 30 lb 4 oz

16.  7 ft 8 in

17.  2 m 27 cm

18.  16.56 lb

19.  7.75 m

20.  152.25 ft

21.  $25 \text{ L} \times \dfrac{1.057 \text{ qt}}{1 \text{ L}} = 26.43 \text{ qt}$

22.  2.07 mi

23.  43.51 mi

24.  2.66 L

25.  2,365.18 cc

26.  274 mm

27.  0.25 mi

28.  54 hr

29.  57,600 sec

30.  77.23 km

31.  $50.4 \text{ ft}^2 \times \dfrac{144 \text{ in}^2}{1 \text{ ft}^2} = 7,257.6 \text{ in}^2$

32.  225,000 mm²

33.  1.32 in²

34.  954.95 ft²

35.  810 ft²

36.  2.35 acres

37.  1,100 cc

38.  3.4 L

39.  53.47 ft³

40.  1,765.5 ft³

41.  6.8 m³

42.  473.12 cc

43.  8.45 fl oz

44.  51,840 in³

45.  262.43 in³

46.  24.94 hr

47.  $\dfrac{\$810}{100 \text{ L}} = \$8.1 / \text{L}$        $\dfrac{\$685.05}{35 \text{ gal}} = \$19.57 / \text{gal}$

Now convert $8.1 / L to $ / gal to compare.

$\dfrac{\$8.1}{1 \text{ L}} \times \dfrac{3.785 \text{ L}}{1 \text{ gal}} = \$30.66 / \text{gal}$

The 35 gal drum is the better buy.

48. 500 copy lot     49. 36,266.67        50. 72.41 km/hr     51. 15.47 kg/cm$^2$

52. 2,727.27 mi/hr  53. 750 mi/hr        54. 13.6 km/L        55. Car A, 0.36 mi/gal

56. 70,294.64 kg/m$^2$                    57. 0.10 lb/in$^2$

58. 240 pages $\times \dfrac{1 \text{ min}}{48 \text{ pages}}$ = 5 min

59. 2.15 hom/week                         60. 1,616.43 thefts/yr

61. \$90              62. 1200 Cal         63. 875 Cal          64. 600 cal

65. 2,190 thefts/yr  66. 5 days           67. 5 hr             68. 4 hr

69. \$47,520         70. \$1.70/lb        71. \$1.12/gal       72. 10 kg bag

73. German supplier                       74. 25 packs         75. \$1.50/kg

76. 150 ticket lot

77. $\dfrac{\$16.50}{1 \text{ gal}} \times \dfrac{1 \text{ gal}}{3.781 \text{ L}}$ = \$4.36 / L

    retail price 1.45 $\times$ \$4.36 = \$6.32

78. \$345 person     79. \$191.40 loss, 30.53% loss

80. 42,900 gal/yr, 49,335 gal             81. \$0.89 profit, 83.96% profit

82. \$1.52 profit, 43.68% profit          83. \$2.74 profit, 219.2% profit

84. \$1.60 profit, 400% profit            85. \$8,215,312.5

86. a) 168  b) \$5,250,000  c) \$5,512,500

## 2.1 Answers

1. Var: A, P, T
   Coe: 2 and 3
   Con: 1

2. Var: R, r, t
   Coe: 2 and –4
   Con: 7

3. Var: F, g, t
   Coe: –4 and –2
   Con: 4

4. Var: C and t
   Coe: –3
   Con: 4

5. Var: B, p, n
   Con: –9

6. Var: D, T
   Con: 24 and 6

7. Var: t and s
   Coe: 2
   Con: 14

8. Var: Z and U

9. $A = 1/2\,b\,h$     10. $C = 2\pi r$     11. $P = 4\,S$     12. $A = \pi r^2$

13. $A = 4\pi r^2$     14. $V = \pi r^2 h$     15. $A = LW$     16. $V = 1/3\,\pi r^2 h$

17. $P = 2W + 2L$     18. $F = 1.8\,C + 32$     19. $P = a + b + c$     20. $V = l\,w\,h$

21. $C = 0.5\,m + 2$     22. $C = 20\,h + 30$     23. $C = 45\,(h - 1) + 60$

24. $N = 10 \bullet 20 \bullet 48\,n$          25. $P = 6\,h + 0.005\,t$

26. $C = 0.08\,c + 26$     27. $C = 720 + 0.02\,c$          28. $C = 32\,d + 0.28\,m$

29. $B = 2 \bullet 4 + 3 \bullet 3 - 5$          30. $T = 400 \bullet 3 - 120 \bullet 5 + 45$
    $B = 12$                    $T = 645$

31. $P = 130$     32. $A = 22.562\ \text{in}^2$     33. $A = 25.7775\ \text{in}^2$  34. $C = \$8$

35. $C = \$80$     36. $C = \$150$     37. $N = 192{,}000$     38. $P = \$400.30$

39. $C = \$200.80$     40. $C = \$1{,}928$     41. $C = \$227.60$     42. $F = 77°$

43. $A = 12.57\ \text{ft}^2$     44. $A = 24\ \text{in}^2$     45. $P = 340\ \text{ft}$     46. Payback = \$1,920

47. $R = \$94.50$     48. Payback = \$15,735.58          49. Payback = \$12,113.75

50. $v = \$803.60$     51. $R = \$328.75$     52. $R = \$5.11$     53. $v = \$341.29$

54. $M = 17{,}404.75\ \text{lb}$          55. $R = \$18.64$     56. $\$1{,}613.13$

57. $\$442$     58. $\$3.75$     59. $\$284.35$

## 2.2 Answers

1. $8x + 8$     2. $7y - 2$     3. $15t + 3q$     4. $2p - r$

5. $-4x - 7$     6. $0$     7. $-x + 3$     8. $3x + 1$

9. $2x^2 + 5x$     10. $8xy - 7x$     11. $x + y$     12. $2a + b - 7c + 7d$

13. $8x^2 - 3x + 9$     14. $13x^2 - 7xy + y^2$          15. $7x^2 - 4x + 6$

16. $6x^2 + 4x + xy + 5y$          17. $45a^2 + 6a + 7$     18. $22t^3 + 7r^2$

19. $4s^3 + 3S^3 + s^2 - 3s - 2$          20. $6p^2 - 9p - 15$     21. $11xy^2 + 7xy + y^2 - 7y$

22. $7pv + pv^2 - p^2 - 2v^2$      23. $10kAT - 3A - 15T$

24. $4mv^3 - 3mv^2 + 12mv$      25. $5R^2 + 4R - 10r^2 - 20r$

26. $14t^2 - t + 3$      27. $52q^2 - 50q - 1$      28. $2.8r^2 + 0.2r + 0.4$

29. $0.5s^2 + 2.36s + 1.56$      30. $3x^2 - 2x$      31. $0.75p^2 + 3p - 0.75$

32. $3r^2 - 2r$      33. $6a^2cp - acp - cp - 3c$      34. $3b^3c^2 + 8b^2c^2 - bc^2$

35. $10xy + 9xy^2$      36. $abc + 2c^2 - 3ac^2$   37. $3s^2 - s + 6$      38. $0$

39. $66y^2 + 64y + 11$

## 2.3 Answers

1. $2x - 3 + 5x + 10$

   $7x + 7$

2. $3x + 12 + 5x - 6$

   $8x + 6$

3. $5x + 8$      4. $-6x + 11$      5. $28x - 8$      6. $-36x + 25$

7. $6x^2 + 12x + 9 - 4x - 5$

   $6x^2 + 8x + 4$

8. $4x^2 - 17x - 2$      9. $4x^2 - 13x + 16$      10. $6a - 5b + 11c$      11. $2x$

12. $19s - 2t$      13. $-3x + 2$      14. $4 - 3x$      15. $0$

16. $x + 3 - x + 3$

   $\phantom{xxx}6$

17. $4x - 7 - 3x + 7$

   $\phantom{xxxxx}x$

18. $4x + 2$      19. $3x - 7$      20. $0$      21. $x^2 + 2x + 2$

22. $5x^2 - 11x + 8$      23. $4x^2 + xy + 3y^2$      24. $11a^2 - b - 5c$      25. $4rs + 3s^2$

26. $0.2x - 3.2y$      27. $1.5x + 4.4$      28. $-x + 3.56y$      29. $-3.4x + 9.85y$

30. $3.4m + 9.9n$      31. $3x^2 - 13.3x + 4$   32. $-6$      33. $-3a^2 + 14b^2 + 2c^3 + 15$

34. $-2y + w$      35. $-1.2x + 0.6y$

36. $4x - 7 - (2x + 5)$

$\quad\ 4x - 7 - 2x - 5$

$\quad\ 2x - 12$

37. $7x - 8 - (4x - 3)$

$\quad\ 7x - 8 - 4x + 3$

$\quad\ 3x - 5$

38. $2x^2 + x + 19$

39. $x^2 - 10x + 9$

40. $10x + 3$

41. $29x + 1$

42. $4x - 34$

43. $14x + 50$

44. $11x^2 + 29x - 9$

45. $2x^2 + 48x - 23$

46. $11x^2 - 6x + 31$

47. $2x^2 + xz + 5z^2$

48. $x^2 + 15x + 6$

49. $-3x^2 + 10x + 25$

50. $33x + 60$

51. $29x - 49$

2.4 Answers

1. $3x^2 - 5x$

2. $8x^2 + 14x$

3. $15x^2 - 6x$

4. $35x^2 - 15x$

5. $2x^3 - 6x^2 - 10x$

6. $12x^3 + 15x^2 - 21x$

7. $12x^3 - 44x^2 - 8x$

8. $-35x^3 + 15x^2 - 80x$

9. $-2x^3 - 6x^2 + 10x + 3x^2 + 6x - 7$

$\quad\ -2x^3 - 3x^2 + 16x - 7$

10. $2x^3 - 2x^2 + 3x^3 + 12x^2 + 6$

$\quad\ 5x^3 + 10x^2 + 6$

11. $39x - 19x^2$

12. $2x^3 + 2x^2 - 4x$

13. $12x^7$

14. $14x^8$

15. $-64x^9$

16. $21x^5$

17. $-15m^6n^5$

18. $-12m^2 n^6$

19. $-6m^8n^3$

20. $13x^3 - 11x^2y$

21. $19r^2 + 5rt - 4t$

22. $2a^4 + 2a^2 b + 4a^2c - 10bc$

23. $-v^3 - 2uv + 3v^2$

24. $6x^3 + 7x^3y + 5x^2y$

25. $x$

26. $-4y$

27. $2x^3$

28. $-7t^3$

29. $-6x^2$

30. $5x^2$

31. $-5xy^2$

32. $-16x^2y^5$

33. 3x + 2x

   5x

34. 3y + 2y

   5y

35. 3y + 6y − 10

   9y − 10

36. −4b −8c + 4c$^2$

37. 4 − ax

38. 5x − y

39. 2

40. −9x$^2$ + 10x

41. 3y$^2$ − 10y

42. 4a − 5b

43. −6r + 10s

44. 2T$_1$$^2$ + 6T$_1$

45. 10R$_1$$^2$ − 17R$_1$R$_2$

46. 14s$_1$$^5$ − 25s$_1$$^4$ + 23s$_1$$^3$

2.5 Answers

1. n + 5

2. 4 + n

3. n − 3

4. 17 − n

5. 13 − 2n

6. 7 + 3n

7. n/2 − 8

8. 4n − 9

9. 5n − 63

10. 3 ( n + 12 )

11. 4 ( n − 15 )

12. 1/5 ( n + 21 )

13. 3 ( 11 − n )

14. 1/6 ( 5n )

15. n$^2$ − 6

16. 6 − 2n

17. n/4 − 6

18. 7 + n$^3$

19. $3\sqrt{n}$

20. $(n - 12)^2$

21. $(n + 8)^2$

22. $4(n + 5)^2$

23. $3(n - 16)^2$

24. $\sqrt{n + 6}$

25. $\sqrt{n - 11}$

26. $6n^2 - 5$

27. $\dfrac{n^2}{4} + 7$

28. $2\left(\dfrac{n}{2} + \dfrac{1}{4}\right)$

29. $4\left(\dfrac{n}{5} + \dfrac{1}{3}\right)$

30. $6\left(\dfrac{3}{5}n - \dfrac{1}{6}\right)$

31. $\dfrac{5}{11}n - \dfrac{3}{5}$

32. $\dfrac{4}{n} + 1$

33. $1 - \dfrac{1}{n}$

34. 6 + 7n

35. $7 + 2n^2$

36. $4n^2 - 11$

37. $n^2 + 3n$

38. $16 - \sqrt{n}$

39. ( n + 6 ) ( n − 4 )

40. ( n + 5 ) ( n + 3 )

41. $6x^2 - 3x + 4x^2$

$\qquad 10x^2 - 3x$

$\qquad 10(-2)^2 - 3(-2)$

$\qquad\qquad 46$

42. $14x^2 - 21xy + 42xy - 3x^2$

$\qquad 11x^2 + 21xy$

$\qquad 11(-1)^2 + 21(-1)(-2)$

$\qquad\qquad 53$

43. –315

44. 635

45. –21

46. 55

47. 42

48. –25

49. 312

50. –128

51. 21

52. 19

53. 24

54. 73

55. –80

56. 36

57. –219

58. –390

59. –20

60. 5

## 2.6 Answers

1.
a) A concise, condensed set of mathematical instructions in equation form.

b) A formula which holds true for any member of a given class. For example, the formula for the area of a rectangle, A = LW, is true for any rectangle.

c) A formula which only works for a particular item, or company product, etc., but not every member of that category.

d) Letters that represent numbers (either constants or variables).

e) Letters which represent values which can change. Example: x in 3x + y = 1.

f) A number which tells us how many of a certain variable we have.

g) A value which never varies.

h) Any quantity made up of products, quotients, powers, and roots of numbers and/or literal symbols. Examples: 3, 9x, $2x^2$, $3x^3y^5$ and $2\pi x$

i) Any combination of sums and differences of algebraic terms.

j) Terms in an algebraic expression that differ only by coefficient.

k) We may only add or subtract like terms.

l) If a, b, and c are any three numbers, then a ( b ± c ) = ab ± ac

m) Applying the distributive rule. Example: –1 ( x + y ) = –x – y. The negative one has been distributed on the right side of the equation.

n) A literal symbol which is raised to some power. Example: $a^m \cdot a^n = a^{m+n}$ "a" is the base.

o) A number written above and to the right of another number, literal symbol, or algebraic term, which tells how many times the term is multiplied by itself. Example: in $x^2$, 2 is the exponent.

p) Same as an exponent.

q) Numbers or literal symbols written small, above and to the right of a term. Usually, these are exponents.

r) Numbers or literal symbols written below, to the right, and small.

274

Example: $X_n$ , n is written in subscript.  It usually denotes one of a series, like in the term ( $X_1 + X_2 + X_3 + \ldots X_n$ ).

s) A term raised to the 2nd power.

t) $\sqrt[n]{x}$ is the nth root of x. $\sqrt{x}$ is the square root of x. The root tells us what number must be multiplied by itself ( and how many times ) to get the term under the root sign.

u) To substitute some specified values into the formula and calculate the desired quantity.

2. Arithmetic involves operations involving only numbers, usually + , − , × , ÷. Algebra involves operations using both numbers and literal symbols, wherein letters may represent variables, constants, or unknown values. Algebra also involves the use of formulas to solve certain types of problems. Arithmetic can be considered part of algebra, but not vice–versa.

3. Var: A, R, S
   Coe: 3 and 5
   Con: 12

4. Var: P, r, t
   Coe: 3 and −4
   Con: 7

5. Var: T, p, q
   Coe: −5 and −2
   Con: 11

6. Var: B and z
   Coe: 6
   Con: 2

7. Var: Q, q, w
   Con: −8

8. Var: D, $T_2$
   Con: 24 and 6

9. Var: h and s
   Coe: 2
   Con: 4

10. Var: G, t, r ,s
    Coe: −3 , 4

11. $P = a + b + c$

12. $C = \pi\, d$

13. $D = st$

14. $r = c \div 2\pi$

15. $C = 5/9\, ( F - 32)$

16. $m = W \div g$

17. $V = 4/3\, \pi\, r^3$

18. $C = 90 \bullet 12 + 0.02c$
    $C = 1080 + 0.02c$

19. $C = 125 + 17.50p$

20. $N = 24\, ( 12n)$
    $N = 24\, ( 12 \bullet 35 )$
    $N = 10{,}080$ boxes

21. $P = 440 + 16.5\, ( t - 40 )$
    $P = 440 + 16.5\, ( 44 - 40 )$
    $P = \$506$

22. $C = 100 + 40\, ( h - 1 )$
    $C = 100 + 40\, ( 4 - 1 )$
    $C = \$220$

23. $C = 75 + 1.25p$
    $C = 75 + 1.25 \bullet 12{,}000$
    $C = \$15{,}075$

24. $R = 3 \bullet 4 + 3 \bullet 3 + 12$
    $R = 33$

25. $D = 120 \bullet 3 - 48 \bullet 5 + 36$
    $D = 156$

26. T = 430

27. P = 10.34 in

28. C = 23.56 cm

29. D = 112.5 mi

30. r = 4 in

31. C = 100°

32. m = 12.5 slugs

33. $V = 268.08 \text{ in}^3$

34. C = $2,889

35. C = $96,375

36. R = 5.29 ohms

37. $A = 855.3 \text{ cm}^2$

38. $V = 1,884.96 \text{ cm}^3$

39. d = 96 ft

40. $A = 1,432.57 \text{ ft}^2$

41. $8x + 8$

42. $9y - 2$

43. $11t + 6q$

44. $5p + 4r$

45. 0

46. $2x^2 + 11x$

47. $4x + 5$

48. $x + 1$

49. $3x^2 - 6x$

50. $8xy - 7x$

51. $20y$

52. $-3x + 3y - 4z + 7t$

53. $13p^2 + 7p + 7$

54. $3y^2 + 15xy - 3x^2$

55. $23t^2 + 10t + 5$

56. $25z^2 + 3z + xz + 5x$

57. $25x^2 + 3x + 23$

58. $22x^3 + 11x^2$

59. $30x^5 - 5x^3 - 57x$

60. $25t^2 - 19t + 22$

61. $21xy^2 + 27xy + 8y^2 - 17y$

62. $8ab + ab^2 + a^2 - 10b^2$

63. $14cAT + 32T + 11A$

64. $9xy^3 + 14xy^2 + 12xy$

65. $15T^2 + 7t^2 + 9T - 20t$

66. $16x^2 + 37x - 14$

67. $32r^2 - 21r - 1$

68. $1.8x^2 + 0.6x + 2.4$

69. $0.1y^2 + 2.66y + 3.15$

70. $x^2 - 6/7 \, x$

71. $5.55x^2 + 1.5x - 1.35$

72. $4.5t^2 - 0.9t$

73. $6x^3 y^2 + 6x^2y^2 + 2xy^2$

74. $16xy + 10xy^2 + 15x^2y$

75. $2r^2 + 7$

76. $152x + 58x^2 - 14$

77. $-x^2 + 45x + 27$

78. $18x - 50y - 43z$

79. $5x - 3 + 2x + 4$
$7x + 1$

80. $2x + 6 + 7x - 5$
$9x + 1$

81. $16x + 25$

82. $-32x + 5$

83. $-11 + 20x$

84. $-60x + 51$

85. $40x^2 + 49x + 60$

86. $-19x^2 - 58x - 61$

87. $-5x^2 - 24x - 4$

88. $22a - 22b + 28c$

89. $12x - 25$

90. $15x + 20y$

91. $-5t + 12$

92. $16 - 5x$

93. 0　　　　　94. 0　　　　　95. $x - 2$　　　　96. 0

97. $4x + 2$　　　98. $-2z + 2$　　　99. $-2x - 1$　　　100. $x^2 + 17x + 8$

101. $13x^2 - 11x + 19$　　　　　　102. $-14x^2 + 25xy + 45y^2$

103. $42a^2 - 30b + 22c$　　　　　104. $5.4x - 9.1y$　　　105. $6.2x + 5.45$

106. $0.25x - 0.45y$　　107. $12.6x + 57y$　　108. $-13.5t + 12.7s$　109. $6.5x^2 - 7.7x + 7.4$

110. $23t^2 - 3t - 17$　　111. $-19x^2 + 3z^3 + 28y^2 + 44$　　　　112. $-7a + 6b$

113. $-0.6t + 1.1s$　　　　　　　114. $4 ( 7x - 9 ) - 3 ( 2x + 5 )$
$$28x - 36 - 6x - 15$$
$$22x - 51$$

115. $3 ( 7x^2 + 5x - 9 ) - 5 ( 3x^2 - 5x - 7 )$　　　　116. $9x^2 + 56x - 8$
$$21x^2 + 15x - 27 - 15x^2 + 25x + 35$$
$$6x^2 + 40x + 8$$

117. $2x^2 - 11x + 21$　118. $8x^2 + 21xy - 4y^2$　　　　119. $2x^2 + 6x + 15$

120. $4x^2 + 23x + 17$　121. $43x - 1$　　122. $-31x - 75$　　123. $4x^2 - 7x$

124. $-9x^2 - 6x$　　　125. $14x^2 - 4x$　　126. $35x^2 - 45x$　　127. $8x^3 - 20x^2 + 28x$

128. $-20x^3 - 28x^2 + 8x$　　　　129. $15x^3 + 21x^2 - 12x$

130. $-10x^3 + 16x^2 + 4x$　　　　131. $-3y^3 - 18y^2 - 19y - 9$

132. $7t^3 - 2t^2 + 6$　　133. $19z + 2z^2$　　134. $20t^3 + 14t^2$　　135. $-32x^{11}$

136. $-14x^9$　　　　137. $-4t^{12}$　　　138. $-40s^7$　　　139. $-10a^5b^5$

140. $-10s^6t^3$　　　141. $29x^2 - 10y + 17xy$

142. $-3x^4 - 5x^2y + 16x^2z + 28yz$　　　143. $x^3 + 17x^3y + 2x^2y$

144. $-5x^4$　　　　145. $-16r$　　　146. $8x^2$　　　147. $10t$

148. $5tr + r - 4$　　149. $6b - 10 + 2a$　　150. $-7y - 12z + 11z^2$

151. $by + 8$　　　152. $18x - 21$　　153. $16x + 30$　　154. $4x + 25$

155. $19R_1$   156. $6T_1^2 + 12T_1T_2$   157. $10a_1^5 + 21a_1^4 - 43a_1^3$

158. $12y^2 - 28y + 5y^2$
    $17y^2 - 28y$
        $45$

159. $-11$   160. $-1,059$

161. $-199$   162. $-13$   163. $-35$   164. $-58$

165. $-1$   166. $958$   167. $100$   168. $-153$

169. $6$   170. $70$   171. $10$   172. $57$

173. $-134$   174. $-144$   175. $n - 4$   176. $n - 5$

177. $n - 3$   178. $12 - n$   179. $14 + 2n$   180. $3n - 6$

181. $2n - 7$   182. $7n - 10$   183. $5n - 63$   184. $5(n - 23)$

185. $6(n + 15)$   186. $1/4(n + 13)$   187. $5(8 - n)$   188. $3n - 8$

189. $1/5(2n)$   190. $7 + n^2$   191. $8 + n/4$   192. $(3 + n) + 7$

193. $(6 - n) - 5$   194. $(n + 11)^2$   195. $9(n - 14)^2$   196. $5(n - 9)^2$

197. $\sqrt{n + 3}$   198. $\sqrt{n - 19}$   199. $4n^2 + 12$   200. $1/7\, n^2 - 7$

201. $6\left(\dfrac{1}{3}n + \dfrac{1}{7}\right)$   202. $\dfrac{1}{2}\left(n + \dfrac{1}{4}\right)$   203. $5(2n - 4)$   204. $11(n + 6)$

205. $10(4 - n) + 7$   206. $16 - 5n$   207. $8 + 3n$   208. $7 + 2n^2$

## 3.1 Answers

1. $3x = 9$
   Dividing each side by 3:
   $x = 3$

2. $t - 1 = 6$
   Adding 1 to both sides:
   $t = 7$

3. $s = 6$   4. $y = 10$   5. $r = 2$   6. $h = -4$

7. $w = 20$   8. $x = 3$   9. $3y - 1 = 8$
   Adding 1 to both sides:
   $3y = 9$
   Dividing each side by 3: $y = 3$

10. $y = 2$   11. $t = 3$   12. $p = 3$   13. $q = -2$

278

14. m = 48          15. z = 36          16. t = 8          17. w = −35

18. y = −2          19. p = 2          20. m − 5m = −16
                                        Adding the left side:
                                        −4m = −16
                                        Dividing both sides by −4:
                                        m = 4

21. t = 5          22. q = −9          23. y = 1          24. x = 2

25. x = 30         26. x = 21          27. t = −18/5      28. p = −20

29. r = 14         30. 9x − 4 = 7x                        31. x = 3
                       Subtracting 7x from both sides:
                       2x − 4 = 0
                       Adding 4 to both sides:
                       2x = 4
                       Dividing each side by 2:
                       x = 2

32. t = 3          33. p = −1          34. y = 10/3       35. n = 7/3

36. t = 2          37. q = −17         38. d = 3          39. r = 1/3

40. p = 1          41. s = 1           42. m = 4/9        43. s = 40

44. s = 50         45. w = 10          46. w = 12         47. x = 40

48. x = 20         49. x = 5000        50. x = 8000       51. y = 5

52. x = 6.67       53. a = 3.73        54. x = 24         55. x = 3.33

56. $\frac{2}{3}x + 4 = 2x - 4$

    Multiplying both sides by 3:

    $2x + 12 = 6x - 12$

    Subtracting both sides by 2x:

    $12 = 4x - 12$

    Adding 12 to both sides:

    $24 = 4x$

    Dividing both sides by 4:

    $6 = x$

57. $x = 14$

58. $x = 12$

59. $x = 20$

60. $x = 2$

61. $x = 70$

62. $3(2x - 1) = 27$

    Distributing on the left side:

    $6x - 3 = 27$

    Adding 3 to both sides:

    $6x = 30$

    Dividing both sides by 6:

    $x = 5$

    conditional

63. conditional

64. contradiction

65. contradiction

66. identical

67. identical

68. contradiction

69. contradiction

70. identical

71. conditional

72. identical

73. a) $T = 20$  b) $V = -941.31$  c) $T_2 = 1,170.25$

74. a) $11x + 5y$  b) $24x^2 - 2x - 12$  c) $4V^2 - 4VR$

## 3.2 Answers

1. $y = 2x - a$

   Adding a to both sides:

   $y + a = 2x$

   Dividing both sides by 2:

   $\frac{y + a}{2} = x$

2. $y = kx + 3$

   Subtracting 3 from both sides:

   $y - 3 = kx$

   Dividing both sides by k:

   $\frac{y - 3}{k} = x$

3. $x = \dfrac{y - c}{b}$    4. $x = \dfrac{y - A_2}{A_1}$   5. $x = \dfrac{y - 3b}{2c}$   6. $x = \dfrac{3 - y}{4kz}$

7. $y = \dfrac{-2k + ax}{3}$    8. $y = \dfrac{3A_1 x - 2A_1}{4A_2}$

9. $x = \dfrac{y + 2c}{5}$   10. $x = \dfrac{5y}{2a - 3}$   11. $y = \dfrac{-3x}{2k - 3}$   12. $x = \dfrac{4y - c}{6 + 3b}$

13. $y = \dfrac{-10x}{A_1 - A_2}$   14. $2a + 10x = y$   15. $y = \dfrac{2x - 8a}{3a}$   16. $y = \dfrac{14x}{b}$

17. $x = \dfrac{-0.28\,mc + 0.3y}{0.58m}$    18. $y = \dfrac{0.8x + 9.34bx}{2.86b}$

19. $st = d$    20. $t = \dfrac{d}{s}$    21. $x - sz = \bar{x}$   22. $P - a - c = b$

23. $d = \dfrac{C}{\pi}$    24. $a = 2S - b - c$    25. $D = \dfrac{B}{MT}$

26. $L = \dfrac{P - 2W}{2}$   27. $C = \dfrac{P - 240}{0.03}$   28. $m = \dfrac{P - C}{C}$   29. $V_2 = \dfrac{P_2 V_1}{P_1}$

30. $m = \dfrac{2K}{v^2}$    31. $n = \dfrac{u\,(100 + v)}{vp}$

32. $C_1 = \dfrac{P - 350 - 0.1C_2}{0.05}$    33. $d_1 = \dfrac{F_2 d_2}{F_1}$   34. $A = \dfrac{F}{P}$

35. $d = \dfrac{W}{F}$    36. $V = \dfrac{m}{d}$    37. $R = \dfrac{P}{B}$    38. $B = \dfrac{P}{R}$

39. $b_1 = \dfrac{2A - b_2 h}{h}$    40. $h = \dfrac{P}{mg}$   41. $p = \dfrac{2A}{a}$

42. $S = \dfrac{gt^2}{2}$    43. $r = \dfrac{I}{pt}$    44. $\pi = \dfrac{C}{d}$    45. $p = \dfrac{A}{1 + rt}$

46. $C = \dfrac{5}{9}\,(F - 32)$    47. $\dfrac{n_2}{n_1}\,V_1 = V_2$    48. $t = \dfrac{W + 20p}{1.5p}$

49. $\pi = \dfrac{4A}{d^2}$    50. $F_2 = \dfrac{F_1 x}{c - x}$    51. $s = c - dn$    52. $h = \dfrac{V}{\pi r^2}$

53. $r = \sqrt{\dfrac{V}{\pi h}}$    54. $r = \dfrac{LA}{\pi h_s}$    55. $t = \dfrac{A - p}{pr}$    56. $r = \sqrt{\dfrac{3V}{\pi h}}$

57. $s = \sqrt{A}$    58. $d = \sqrt{\dfrac{4A}{\pi}}$    59. $v = \sqrt{\dfrac{2K}{m}}$    60. $B = \dfrac{180 - A}{2}$

61. $R = \dfrac{360I}{PT}$    62. $c = \dfrac{2b}{a}$    63. $b = 3\bar{x} - a - c$

64. $T_1 = \dfrac{kAT_2 - QL}{kA}$    65. $a = \dfrac{v - v_o}{t}$    66. $\pi = \dfrac{V}{r^2 h}$

67. $r = \sqrt{\dfrac{V}{\pi h}}$    68. $r = \dfrac{A - p}{pt}$    69. $c = \dfrac{d + rs}{r}$    70. $t = \dfrac{m - p}{pr}$

71. $m = \dfrac{p}{1 - dt}$    72. $d = \dfrac{m - p}{mt}$    73. $R = \dfrac{I}{PT}$    74. $T = \dfrac{PV}{nR}$

75. $C = S - M$    76. $V = \dfrac{nRT}{P}$    77. $R = \dfrac{s^2}{14.9769f}$

78. $D = \dfrac{s^2}{30F}$    79. $D = \dfrac{s^2}{30f}$    80. $C = \sqrt{8MR - 4M^2}$

81. $n - 7$    82. $n - 8$    83. $10 - n$    84. $2n - 5$

85. $10 + 3n$    86. $4(n + 7)$    87. $5(n - 10)$    88. $3(9 - n)$

89. $9 + n^2$       90. $n/2 - 4$       91. $2n - 7$       92. $(n + 7)^2$

93. $2(n - 2)^2$       94. $4n^2 + 12$       95. $3/7(n + 14)$    96. $3(3n - 5)$

## 3.3 Answers

1.  $P = BR$
    $P$ = commission earned, $B$ = total cost of trip, $R$ = percent received.
    $275 = 3{,}600R$
    Dividing both sides by 3,600:
    $0.076 = R$
    The agent received 7.6%.

2.  $P = BR$
    $P$ = amount spent on advertising, $B$ = total sales, $R$ = percent of sales.
    $P = 9{,}275 (0.047)$
    Multiplying on right hand side:
    $P = \$435.93$
    $435.93 was spent on advertising.

| | | | |
|---|---|---|---|
| 3. $2,000 | 4. 14.3% | 5. 8% | 6. $76.92 |
| 7. $2,231 | 8. $38,000 | 9. 4.4% | 10. $10,950 |
| 11. 15.7% | 12. $9 | 13. 6% | 14. 41,907 |
| 15. $4,153.85 | 16. $1,200 | 17. $203,000 | 18. 137 |
| 19. 9.6 hr | 20. 11.8% | 21. $126,000 | 22. 101 rooms |
| 23. 6.5 lb | 24. 8.5% | 25. $152,173.91 | 26. $1,050 |
| 27. 15.8% | 28. 16% | 29. 60% | 30. 3,125 |

## 3.4 Answers

1.  Let $n$ = the number
    $2(n - 7) = 46$
    $2n - 14 = 46$
    $2n = 60$
    $n = 30$
    Verify: $2(n - 7) = 2(30 - 7) = 2 \bullet 23 = 46$

2. Let n = the second number. Then 9n + 8 is the first number. Since the two numbers add to 88, we have:

n + 9n + 8 = 88

10n = 80

n = 8

Then the first number is 9n + 8 = 9 • 8 + 8 = 80

Verify: 80 + 8 = 88

3. 20, 53          4. 15, 43, 33          5. 112, 114, 116     6. 10 years, 25 years

7. 6 ft, 10 ft          8. 6 years, 30 years 9. $22.80, $17.20

10. $90.45 floor lamp, $62.30 desk lamp 11. $0.60, $0.50     12. 8,326.5 min of calls

13. new clerk 12 pkgs/hr, middle clerk 17 pkgs/hr, senior clerk 20 pkgs/hr

14. division I $21,473,684.21, division II $7,157,894.74, division III $5,368,421.05

15. 3 hours          16. 2.5 hours          17. 1,803 tickets     18. 10 twentys

19. 150 quarters, 132 half dollars, 190 dimes          20. 50 L

21. 24 gal          22. 20          23. a) 12 mi/hr  b) 195 km  c) 5.8 hr

24. Let x = time          25. 2 hours          26. 2:42 p.m.

12x + 8x = 1,000

20x = 1,000

x = 50 hours

Verify: 12 • 50 + 8 • 50 = 1,000

1,000 = 1,000

27. 1,440 mi          28. 50 mi/hr          29. 360 mi          30. 21.6 hr

31. 4 hours          32. 10:15 a.m.

33. Let x = the weight of 10% beans. Then 20 − x is the weight of 50% beans.

0.5 ( 20 − x ) + 0.1x = 0.2 ( 20 )

10 − 0.5x + 0.1x = 4

10 − 0.4x = 4

−0.4x = −6

x = 15 lb of 10% beans

Then 20 − x = 20 − 15 = 5 lb of 50% beans.

Verify: 0.1 • 15 + 0.5 • 5 = 4 which is 20% of 20.

34. 1.8 L  35. 0.824 gal  36. 30 lb at $5.00, 20 lb at $2.50

37. 5.71 kg  38. 2.3 kg  39. 13.33 gal

40. Let x = amount invested at 9%. Then 10,000 − x = amount invested at 7%.
    0.09x + 0.07 ( 10,000 − x ) = 830
    0.09x + 700 − 0.07x = 830
    0.02x = 130
    x = $6,500 at 9%
    Then 10,000 − x = 10,000 − 6,500 = $3,500 at 7%.
    Verify:  0.09 • 6,500 + 0.07 • 3,500 = $830.

41. $2,000,000 at 12%, $1,500,000 at 7% loss  42. $24,106.70

43. $21,526.17  44. $10,000 and $25,000

45. $3,500,000 and $3,500,000  46. $29,015.75  47. $18,103.45

3.5 Answers

1. 13/5  2. 3/1  3. 3/7  4. 9/5

5. 5/3  6. 12/7  7. 3/11  8. 5/1

9. 7/6  10. 3/2  11. 4/3  12. 4/3

13. 3/1  14. 1/7

15. $\dfrac{2\text{ m}}{40\text{ cm}} = \dfrac{200\text{ cm}}{40\text{ cm}} = \dfrac{5}{1}$  16. $\dfrac{1.5\text{ lb}}{4\text{ oz}} = \dfrac{1.5\text{ lb}}{0.25\text{ lb}} = \dfrac{6}{1}$

17. 8 lb/3 in$^2$  18. 0.21 ohms/25 ft  19. 2 in/7 lb

20. 1 L/15 sec  21. 12/1  22. 1/6  23. 2/9

24. 24 lb/1 ft$^3$  25. 8000/1  26. 0.25 in/2  27. 88 ft/1 sec

28. 748.35 g/36.32 cm$^3$  29. 1,200 ft$^3$/52 min

30. 165/74 = 2.2 Mach  31. 1 gal/22.3 mi  32. 32.8 lb/1.3 in$^2$

33. 165.5/62  34. 16/5  35. x = 9  36. x = 126

37. x = 60          38. x = 15          39. x = 14          40. x = 3

41. x = 2,400       42. x = 50

43. $\dfrac{1{,}620 \text{ ft}}{2 \text{ sec}} = \dfrac{x \text{ ft}}{27 \text{ sec}}$          44. $\dfrac{21 \text{ fulltime}}{10{,}000 \text{ citizens}} = \dfrac{x \text{ fulltime}}{27{,}500 \text{ citizens}}$

$$1{,}620 \bullet 27 = 2x \qquad\qquad 21 \bullet 27{,}500 = 10{,}000x$$
$$43{,}740 = 2x \qquad\qquad 577{,}500 = 10{,}000x$$
$$21{,}870 \text{ ft} = x \qquad\qquad 57.75 \text{ fulltime} = x$$

45. 2,050 parts     46. $1,335.60     47. 8.5 min       48. 2.5 oz

49. 0.33 min        50. 1,050 part time   51. 15.75 min   52. 2,372.5 inmates

53. 5.25 hr         54. 1,961 female   55. 12,000 acres   56. 21 min

57. 80 pieces       58. 181.74 cases   59. 183,900 portable

60. 180 double      61. 2.34 ft³       62. 33.75 in       63. $4,000,000 , $10,000,000

64. 1,774.2 copies  65. 54,000 don't   66. 20 ft          67. 412.5 mi

68. $25,000         69. 10 years

## 3.6 Answers

1.  a) A statement that two quantities are equal.
    b) A statement that two algebraic expressions are equal.
    c) The highest sum of the powers of the variables in any one term of the equation.
    d) Any value which, when substituted into the equation, yields a true statement.
    e) Isolating a variable in an equation.
    f) An equation which is only true for certain values of its variable.
    g) True for any value of its variable.
    h) Never true for any value of its variable.
    i) An equation in which some or all of the constants are represented by letters.
    j) A literal equation relating two or more physical quantities.
    k) The first quantity divided by a second quantity.
    l) A statement of equality between two ratios.

2. x = 10          3. s = –6          4. t = 20          5. x = –1

6. x = –4          7. s = 3           8. t = 2           9. x = 3

10. x = 1          11. y = 12         12. z = 2          13. x = 2

14. x = 4          15. y = 4          16. x = 4          17. x = 2

18. x = 68.49      19. x = 6.67       20. x = 28         21. x = 16.67

22. x = 10         23. x = 8          24. x = –2         25. x = 3

26. x = 1/2        27. conditional    28. conditional    29. contradiction

30. identical      31. identical      32. conditional    33. contradiction

34. identical

35. $x = \dfrac{y + b}{3}$    36. $x = \dfrac{7 - y}{a}$    37. $x = \dfrac{y + 4k}{7b}$    38. $z = \dfrac{y - 18}{12c}$

39. $y = \dfrac{2ax + 2}{7b}$    40. $y = \dfrac{3cx - 2}{4b}$    41. $x = \dfrac{-y - 6y}{3 + 5b}$    42. $y = \dfrac{7 - 3x}{3 - 2k}$

43. $x = \dfrac{3y - 18b}{21}$    44. $y = \dfrac{36a - 3x}{5a}$    45. $y = \dfrac{0.5ax}{0.16a - 0.3}$

46. $y = \dfrac{8.02x}{14.35}$    47. $h = \dfrac{A}{b}$    48. $W = \dfrac{P - 2L}{2}$    49. $t = \dfrac{P - p}{pr}$

50. $C = \dfrac{P}{1 + m}$    51. $c = \dfrac{T + 3b}{3}$    52. $F_2 = \dfrac{F_1 d_1}{d_2}$    53. $B = \dfrac{P}{R}$

54. $h = \dfrac{2A}{b_1 + b_2}$    55. $m = \dfrac{P}{gh}$    56. $p = \dfrac{A}{1 + rt}$    57. $n_2 = \dfrac{n_1 V_2}{V_1}$

58. $r = \sqrt{\dfrac{A}{\pi}}$    59. $R = \dfrac{yd}{ml}$    60. $t = \sqrt{\dfrac{2S}{g}}$    61. $T_2 = \dfrac{QL + kAT_1}{kA}$

62. $d = \sqrt{\dfrac{k(M_2 - M_1)}{4F}}$

63. 76, 33, 10       64. 35 yr, 42 yr, 12 yr        65. Div B 5 yr
                                                       Div A 15 yr

66. Pants $16.87, Shirts $12.67, Hats $6.77       67. 9,900 min

68. 7.5 hr        69. 500 gal        70. 620 lb        71. 56, 57, and 58

72. 9 ft and 17 ft     73. 2 hr        74. 9 hr

75. 800 mi        76. 55 mph, 45 mph        77. 107.14 hr

78. $3.62, $4.85     79. 2 L        80. 6.67 lb

81. $3,000 at 8%, $2,000 at 3% loss        82. 363.75 gal        83. 2.8 min

84.  20 cm          85.  2.85 mm          86.  3.33 min

87.  420,512 cars, 1,471,792 trucks          88.  3.625 ft          89.  $30,000, $135,000

90.  14.5 ft          91.  $85,000

4.1 Answers

1. – 12.

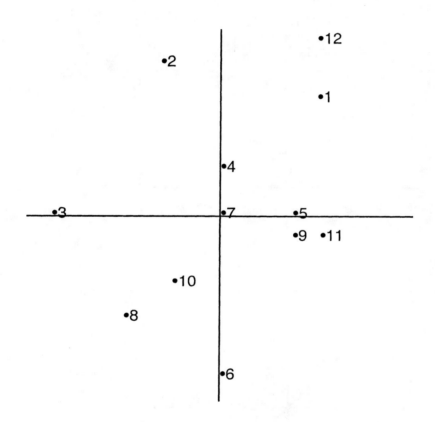

13.  I          14.  II          15.  III          16.  IV

17.  0          18.  0          19.  IV          20.  II

21.  III          22.  I and III          23.  II and IV

## 4.2 Answers

1.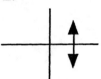

2.

3.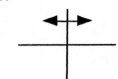

4.

5.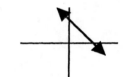

6.

7.

8.

9.

10.

11.

12.

13.

14.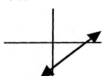

15.

16.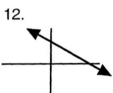

17.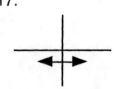

18.

19.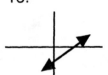

20.

21.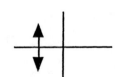

21. $y = 40 - 2x$

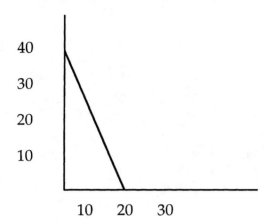

22. $y = 10 - 1.5x$

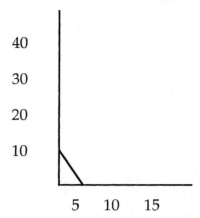

23. a) $20,000        b) $14,000        c) $10,000
    d)

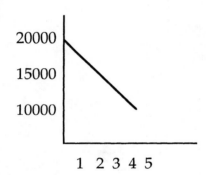

4.3 Answers

1. $m = \dfrac{y_2 - y_1}{x_2 - x_1} = \dfrac{3 - 2}{3 - 5} = \dfrac{1}{-2}$       2. $m = \dfrac{3 - 3}{2 - (-4)} = \dfrac{0}{6} = 0$

3. 4/–3            4. 1            5. undefined        6. undefined

7. 0               8. –3/4        9. 10/11            10. $y - y_1 = m(x - x_1)$
                                                          $y - 3 = 4(x - 1)$
                                                          $y - 3 = 4x - 4$
                                                          $y = 4x - 1$

11. $y = -2x - 1$      12. $y = 3/5\,x + 29/5$                    13. $y = 3$

14. $y = -5/8\,x - 23/8$              15. $x = 1$

291

16. $m = \dfrac{3 - 2}{3 - 5} = \dfrac{-1}{2}$

    $y = \dfrac{-1}{2}x + \dfrac{9}{2}$

17. $y = 3$

18. $y = -4/3\,x + 3$

19. $y = x + 1$

20. $x = 6$

21. $x = 0$

22. $y = 0$

23. $y = -3/4\,x - 7/2$

24. $y = 10/11\,x + 27/11$

25. $3x - 2y = 6$

   Set $x = 0$: $\;3 \bullet 0 - 2y = 6$
                 $-2y = 6$
                    $y = -3$
   So our y–intercept is $( 0, -3 )$

   Set $y = 0$: $\;3x - 2 \bullet 0 = 6$
                   $3x = 6$
                    $x = 2$
   So our x–intercept is $( 2, 0 )$

26. x–intercept is $( 3, 0 )$
    y–intercept is $( 0, 3/2 )$

27. x–intercept $( 4, 0 )$
    y–intercept $( 0, 3 )$

28. x–intercept $( 2, 0 )$
    y–intercept $( 0, -4 )$

## 4.4 Answers

1. $\dfrac{55 - 50}{60 - 50} = \dfrac{x - 5.48}{6.64 - 5.48}$

   $\dfrac{5}{10} = \dfrac{x - 5.48}{1.16}$

   $5.8 = 10x - 54.8$

   $60.6 = 10x$

   $\$6.06 = x$

2. $2.93

3. $28.92

4. 32.5 yr

5. 45 yr

6. 47.5 yr

7. $i = 15a$

8. $375

9. $660

10. $a = i/15$

11. 38 yr

12. 22 yr

13. 65.5 lb

14. 113.7 lb

15. 46.3 in

16. yes, 6 lb

### 4.5 Answers

1. C = dependent
   m = independent

2. C = dependent
   h = independent

3. C = dependent
   m = independent

4. C = dependent
   n = independent

5. y = dependent
   x = independent

6. P = dependent
   r = independent

7. q = dependent
   t = independent

8. y = dependent
   x = independent

9. w = dependent
   z = independent

10. z = dependent
    x = independent

11. C = 10.00 + 0.18m
    C = 10 + 0.18 • 240
    C = $53.20

12. C = $460

13. C = $16.15

14. C = $24,150

15. y = −8

16. P = 24

17. q = 27.33

18. y = 9

19. w = 5

20. z = 570

4.6 Answers

1.

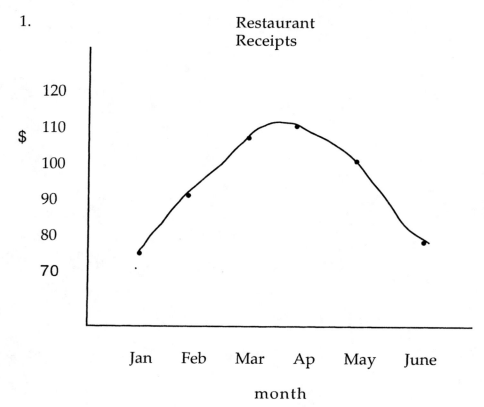

Restaurant Receipts

$

120
110
100
90
80
70

Jan    Feb    Mar    Ap    May    June

month

2.    Gross Receipts

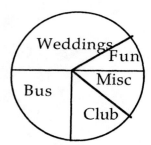

Weddings
Fun
Misc
Bus
Club

3.

Restaurant
Receipts

month

4.

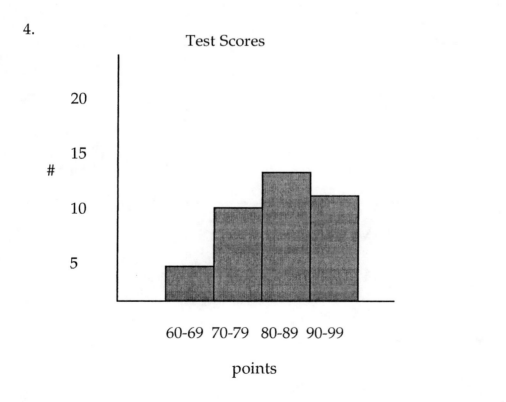

Test Scores

points

295

5.        Staff

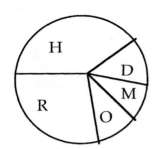

6.                        B & E

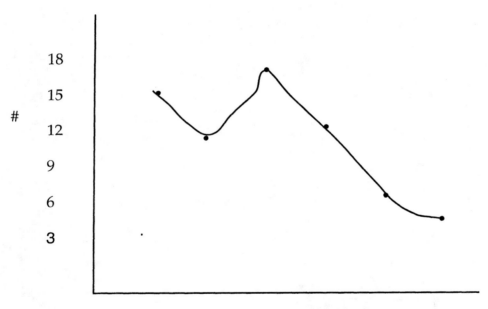

7.

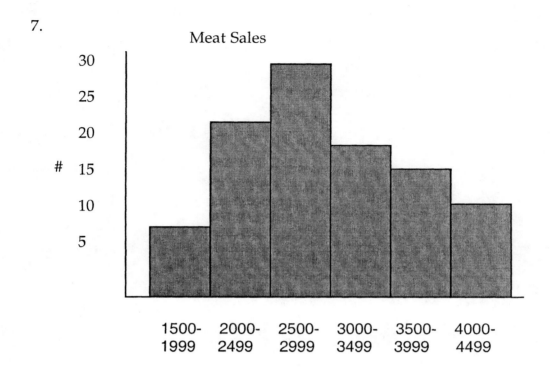

Meat Sales

# 

weights

8.

Average Income

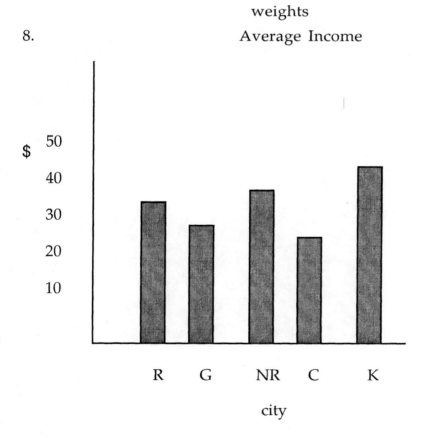

$

city